Praise for *85 Days*

"Witcover has powerfully managed to evoke the passion, the possibilities, the frustrations, the joys and the sadnesses of Robert Kennedy's strange attempt at the Democratic nomination for President . . . *a rare and enormously readable feat of overnight reportage.*"

—*New York Times*

"*Witcover, a Washington correspondent . . . has produced an admirably balanced, temperate, comprehensive account.* He makes clear Kennedy's occasional lapses into demagoguery and expediency as well as his brilliant soaring moments, his moral passion and his private humor."

—*Chicago Tribune*

"Witcover allots sixty pages to the assassination—which he witnessed—and its aftermath. The first six of these pages recreate a single moment of paralyzing terror: the struggle for a gun that waves wildly as five more victims fall, the screams and curses and blood, the photographers jostling to get near the fallen Senator. *It is great reporting, and it is almost unbearable.*"

—*Saturday Review*

"*Witcover's is a fine book, not just because it narrates in rapidly moving fashion a great drama, but because it sketches a blueprint of modern presidential politics.* Even in Kennedy's catch-up, Greek-tragedy way, this is how it was done."

—*Boston Sunday Globe*

Also by Jules Witcover

The American Vice Presidency: From Irrelevance to Power

Joe Biden: A Life of Trial and Redemption

Very Strange Bedfellows: The Short and Unhappy Marriage of Nixon & Agnew

The Making of an Ink-Stained Wretch: Half a Century Pounding the Political Beat

Party of the People: A History of the Democrats

*No Way to Pick a President: How Money and Hired
Guns Have Debased American Elections*

The Year the Dream Died: Revisiting 1968 in America

Mad as Hell: Revolt at the Ballot Box 1992 (with Jack W. Germond)

Crapshoot: Rolling the Dice on the Vice Presidency

*Whose Broad Stripes and Bright Stars?: The Trivial Pursuit
of the Presidency 1988* (with Jack W. Germond)

Wake Us When It's Over: Presidential Politics of 1984 (with Jack W. Germond)

*Blue Smoke & Mirrors: How Reagan Won and Why Carter
Lost the Election of 1980* (with Jack W. Germond)

The Main Change: A Novel

Marathon: The Pursuit of the Presidency 1972–1976

*A Heartbeat Away: The Investigation and Resignation of Vice
President Spiro T. Agnew* (with Richard M. Cohen)

White Knight: The Rise of Spiro Agnew

85 Days

The Last Campaign of Robert Kennedy

JULES WITCOVER

WILLIAM MORROW
An Imprint of HarperCollins*Publishers*

Library of Congress Cataloging-in-Publication Data has been applied for.

ISBN 978-0-06-246391-3

16   17   18   19   20   DIX/RRD   10   9   8   7   6   5   4   3   2   1

In memory of my father

Contents

There is a tide in the affairs of men
Which, taken at the flood, leads on to fortune;
Omitted, all the voyage of their life
Is bound in shallows and in miseries.
On such a full sea are we now afloat;
And we must take the current when it serves,
Or lose our ventures.

JULIUS CAESAR, ACT IV, SCENE III

Foreword

by Senator Edward M. Kennedy

THE YEAR 1968 was not only a year of tragedy for the Kennedy family, it was also a watershed year in contemporary American history. I often think of the "might have beens" of 1968. If Robert Kennedy had lived, he might well have been elected President of the United States, and our entire subsequent history would have been different and better. We might have been spared much of the divisiveness and confrontation that have plagued the nation since then.

I also remember 1968 as a time of great energy and hope, when a majority of the American people first realized that the Vietnam War was wrong, and that it ought to end. Initially, Robert Kennedy had been reluctant to run for president, fearing that his presence in the campaign would divert attention from the issue of the war. He believed that on the war, as on other vital issues of foreign and domestic policy, individuals would make the difference; as he told the students at the University of Capetown in South Africa on his visit there in 1966: "Each time a man stands up for an ideal, or acts to improve the lot of others, or strikes out against injustice, he sends forth a tiny ripple of hope, and crossing each other from a million different centers of energy and daring, those ripples build a current which can sweep down the mightiest walls of oppression and resistance."

It took the better part of a decade after that for America to live up to its ideals and extricate itself from the Vietnam War, but 1968 was the turning point, and Robert Kennedy's campaign was part of the turning. This volume, by one of America's premier journalists, is a vivid chronicle that makes his last campaign come alive again.

In the two decades since Robert Kennedy was taken from us, much has been written about him—by scholars and reporters, classmates in school and college, teammates on the playing fields, and friends from his years of public service. They have told of his achievements; of his character, strength, and faith; of his compassion, humor, and love; of his commitment to others and his vision of America. I knew him in each of these ways, but most of all, of course, I knew him as a brother.

After Joe was killed in World War II, Jack wrote that if the Kennedy family ever amounted to much, it would be because of the leadership Joe had provided. The younger members of our family—and I was the youngest of nine—felt that Bobby set the same kind of example for us.

He always kept his eye on me. When I was ten years old at Fessenden School and he was sixteen at Milton Academy, he would call to find out how my studies were going, and how I was behaving. Occasionally, on fall weekends, we would drive to Cape Cod with our friend Lem Billings and spend the night in the garage of the boarded-up summer house, cooking our own food and sleeping on cots. It is one of my fondest memories in growing up.

On the day before the Harvard-Yale football game at New Haven in November 1953, the Harvard houses were scheduled to play the Yale colleges, and my house, Winthrop, was to play Davenport. Bobby was then a young lawyer on the staff of the Hoover Commission, and he happened to be in New York City working on a report. When I called to ask if he could come and watch the game, he said, "I would rather play." He drove to New Haven and changed into a borrowed uniform in the back of his car. He arrived unnoticed on the field just before the kickoff, and after making several tackles and pass receptions for Winthrop House, we were well ahead of Davenport, and he decided he was not needed any longer as my substitute. He changed again in his car and drove back to New York. No one, including the master of Winthrop House, realized that the "mystery" substitute was Bobby.

Once, when we were serving together in the Senate, he rushed from a meeting with Japanese students to the Senate floor to cast a vote. I had been following the debate, and he looked to me for guidance. He shook his head from side to side, and I nodded my head up and down to indicate that he should vote "Yes." When our names were called, I voted "Yes," but Bobby voted "No." There was laughter in the public gallery as visitors wondered why the Kennedy brothers had disagreed. After the vote was over, he sent me a note which said, "When I shake my head from side to side and you nod your head up and down, does this mean I am to vote 'No' and you agree? Obviously not. In other words, when I wanted to vote 'No,' I should have nodded my head up and down. You would have shaken your head from side to side; then I would have known just how to vote."

Nothing in life mattered more to him than how his children would develop and what kind of lives they would lead. His relationships in the family were his greatest source of strength. From our father, he gained the strong will and toughmindedness that led some to see him as "ruthless"; from our mother, he gained a tender heart and gentle spirit; from our older brothers, he gained a competitive spirit and desire for excellence; and from our sisters, he learned the charm and warmth and shyness that were so appealing, not only to his closest friends but to thousands of others who came to know him in public life or saw him on the campaign trail.

He lived his faith and his religion, and he came to view public service as his calling to help the disadvantaged, the poor, and the victims of discrimination. He was totally involved and committed to their cause. He lived for his children and for the future generations of the earth. He urged others to join him in seeking a newer world, and people responded as perhaps to no other public figure of our time.

His public legacy is written in the achievements of those whom he inspired and who have carried on his work. And there is also the private legacy, the memories of those who knew him best, and miss him most.

I miss the Sunday afternoons at Hickory Hill, when we played touch football with his children, or tennis with General Maxwell Taylor and David Hackett. I miss the ski trips in the West, particularly the early morning milk run, as we rode the chairlift together to take advantage of the fresh powder from an overnight snowfall; once, after a long day on the mountain, we came back to the sauna and rolled in the snow—and he locked me outside at twenty degrees below zero. I miss the sailing trips off the Maine coast; more than once, we were lost in fog or storms, but he always managed to bring us safely into port.

Most of all, I miss our walks across the Capitol grounds from the Senate floor to the Senate office building in every kind of weather. He would urge me to speak at morning assembly at a local high school in Washington to encourage the students to stay in school and continue their education. He would ask me to round up more support for a fund-raiser in Boston for Cesar Chavez's farm workers. He would describe his trip to the Mississippi Delta and the unacceptable hunger, malnutrition, and poverty he saw there. He would talk about American Indians and the injustices they were suffering. These sudden moments of passing conversation, as much as any speech he made, revealed his passionate concern for the forgotten American, and his consuming desire to make things better. As he liked to say, in a phrase of George Bernard Shaw's that became one of the themes of his last campaign, "Some men see things as they are, and say why; I dream things that never were, and say why not?"

When I think of Bobby, I see Cape Cod on a sunny August day. The wind is coming from the southwest, the whitecaps are rising, and the full tide is sweeping through the gaps in the breakwater. It is after lunch, Bobby is stripped to the waist and he is saying, "Come on Kathleen, Joe, Bobby, and David, Courtney, Kerry, come on Michael, Chris, and Max—call your mother and come for a sail." One of the children says, "What about the baby?" And he shouts, "Dougie can come next year." We push off from the landing. The sail of the *Resolute* catches the wind, the boat begins to heel, and there are squeals of delight. Bobby says, "I think today is the day we'll tip over," and the squeals turn to terror as the *Resolute* rounds the end of the breakwater. Beyond, the bow rises and falls with the rhythm of the sea, and the children are covered with salt spray; Bobby dives overboard and catches the line that trails behind, calling to the children to join him. Child after child jumps into the water, grabbing for the line, and those who miss are pulled to it by his strong and suntanned arms. He is the catcher in the sea. The boat heads into Nantucket Sound. The tide is gentle, the gulls are watching from above, the breeze is warm, and there is the happiness and love of being together again.

85 Days

CHAPTER ONE

Fateful Delay

THE DECISION NOT TO RUN

ON THE MORNING of January 30, 1968, Senator Robert F. Kennedy of New York was sitting at a breakfast table in the private President's Room at the National Press Club, answering probing questions from about fifteen political writers and columnists. It was one of those subsurface meetings that have become part of the institutional life of Washington, in which a prominent public figure speaks candidly under a protective arrangement that permits direct quotation only with his approval and limits the rest strictly for the newsmen's "background." The pertinent question had just been asked: were there any circumstances under which he would be a candidate for the Presidency in 1968 against Lyndon B. Johnson?

"No," Senator Kennedy said, "I can't conceive of any circumstances."

A reporter began to suggest some possibilities. Maybe, after five years in the White House, President Johnson would not seek reelection.

"You're talking," Kennedy interjected, "about an act of God."

From this remark, most of those sitting around the private breakfast table concluded that it would take that, and nothing less, to draw Robert Francis Kennedy into direct combat with Lyndon Johnson for the Presidency in an election year that already was one month spent. His own candidacy, he told the group, would tear the party apart, probably resulting in the election of Richard Nixon and a Republican House that would endanger worthwhile national programs in which he believed.

"If I ran I'd have to run in all the primaries," he said. "I don't think I could win the nomination. I would have to win every primary." While he would stand a "fair chance" in each one as it came along, he said, sweeping all of them would be tough. He likened the situation to a horse who might win any given race, but who still would not be likely to be first in a string of them.

But having said this, Robert Kennedy clearly was a man in turmoil. During the backgrounder, this fact surfaced unmistakably in his expressions of deep concern about the Vietnam War and about the alienation abroad in the land— the disquiet among whites and Negroes, the young and elderly, the wealthy and poor. "The cause hasn't been analyzed and dealt with," he said. "There's affluence, yet a feeling of unhappiness in the country. If someone touched the heart of that, and how to bring the country back together—if he could bind the wounds, appeal to the generous nature of Americans. . . ."

What the reporters around the breakfast table were hearing, of course, was the framework of what Robert Kennedy's own campaign would be if he were to run. But in all the compassionate words, there was an overriding sense of futility, of personal helplessness. "If I thought there was anything I could do about it," Kennedy said near the end, "I would do it." He was trying to persuade them that there wasn't, but at the same time he was trying to persuade himself. As one of the witnesses said later, "It was like seeing a man do battle with himself right there before your eyes." After one of his longer explanations for not running, Kennedy concluded: "I know that isn't very satisfactory. But I don't see what I can do." And as the backgrounder broke up, and he shook hands with his hosts, he asked of more than one of them: "Am I doing the right thing?" To some, he murmured, "I know what everybody is saying"—that when the chips were down, he lacked the courage of his convictions.

For Robert Kennedy, whose whole life had been a series of strong positive responses to personal, physical and political challenge, it was the greatest irony to find himself thus in the grip of such uncharacteristic uncertainty and inaction. But forces and circumstances that had been building all through 1967, causing him both political and emotional torment, now were restricting his freedom of action.

By the fall of 1967, two issues were dominating public concern and feeding a growing mood of disaffection in the United States. The first was the war in Vietnam—mired in stalemate, consuming American men and money at an ever-spiraling rate, casting a pall of pessimism and gloom over the conduct of public affairs in every area. The second was the depth of racial unrest at home— the big-city riots, the escalating militancy of the Negro revolt, the consequent demand among fearful whites for stiffer repression and a diminishing interest among them in the removal of racial injustices at the root of the ferment.

Central to these two causes of national disenchantment was the personality of Lyndon Johnson, whose conduct of office, whose style, appearance and excesses drew the lightning of all the dissent. Johnson had contributed to the national climate by his own peculiar brand of dissembling—raising the American stakes in Vietnam by a series of camouflaged intensifications of the conflict; voicing hope for military victory and willingness to talk peace at times when neither was present; promising the eradication of poverty and injustice at home when neither was a realistic possibility in light of federal programs and public attitudes then holding sway.

The two major dilemmas facing the United States, one abroad and one at home, were not of course of Johnson's making, and for all his deficiencies it could not be said that he had not grappled mightily with them. But he had been elected in 1964 as a kind of super-doer, and he was not getting done those things that most needed to be done. Unlike his two immediate predecessors in the White House, Dwight D. Eisenhower and John F. Kennedy, Lyndon Johnson could not count on any storehouse of public goodwill to see him through adversity. He was not, as one of his closest advisers was said once to have told him point-blank, "a likable person." When things went wrong, his image as an insensitive head-knocker impervious to the niceties of personal or diplomatic behavior invariably generated the worst possible public response.

And on the two major issues of the war and urban unrest, and inevitably in the matter of the deep suspicion and animosity toward the occupant of the White House, it was Robert Kennedy who stood as the unchallenged touchstone of the national dissent within the Democratic Party. Not only was he the brother, confidant and political heir of a martyred President and the emotional link to the recent past of taste and elegance in the conduct of national affairs; he also was a bridge to the new America of concerned, involved youth who were particularly disenchanted with the war, the bigotry and Lyndon Johnson. In that period in the months of September and October, those who wanted to change things—and had any pragmatic hope of doing so—had no one else to look to but Robert Kennedy, then in his forty-second year of an eventful and controversial life.

The leading elements of political activism in the dissent—college students and faculty talking of a new politics of participation, well-educated middle-class Americans still embracing the old liberalism, all but the most extremist Negro leaders—saw him as the one man who could chart a new national course. Some valued him chiefly for his electability—the late, deified President's brother whose own ascendancy was somehow preordained, and whose destiny was clearly written in the huge and often frenzied crowds that turned out to see, hear and touch him. There was always the incredible compulsion to touch him—perhaps to take hold of a moment in history or at least to brush against it, or to communicate the compassion that went out from one human being to another in a mutual personal and national loss whose pain lingered.

But there also were others, the campus crowd especially, and the Negroes and white immigrant minorities, who believed they recognized in Robert Kennedy an empathy for their disillusionment with American life. For these groups, the trouble was keyed to Vietnam and to the racial crisis at home, and for many of them the two were inseparable. So it was with Kennedy too. Accordingly, he was better positioned than any other man in public life to forge from these elements and their concerns a potent political coalition that could challenge—and maybe even bring down—the tradition and the power that were supposed to assure an incumbent President renomination by his party.

Kennedy, since his election to the United States Senate from New York in November, 1964, had earned this role as the political embodiment of the nation's dissatisfaction. He was the first politician with serious aspirations and chances for the Presidency to rise in the Senate and broach the unthinkable: that the National Liberation Front would have to be part of any negotiation to end the Vietnam War. He observed on February 19, 1966, that the United States would have to start thinking about the possibility of a coalition government in Saigon in which the NLF would play some role. The idea, at the time Kennedy first expressed it, was written down by many as the greatest political faux pas of his career. Indeed, the public mood had not begun to reject the jingoism that the Johnson Administration had spread like a protective coating over all it did in Vietnam, and the general disposition, except among the early Vietnam critics, was to agree with Vice President Hubert H. Humphrey's likening of the Kennedy notion to letting "a fox in the chicken coop." Throughout 1966 and 1967, Kennedy continued to hound the Administration about its refusal to halt the bombing of North Vietnam and to enter into peace talks. In other areas too—delay in achieving a treaty against nuclear arms proliferation, dealing with Communist China, the decay of the Alliance for Progress in Latin America, the failure to cope with hard-core poverty in the big-city ghettos and in Appalachia—Kennedy hammered away at the executive branch controlled by his own party.

Thus, as the 1968 Presidential election year approached, Kennedy had ample grounds and credentials, in terms of the major issues, for contesting Johnson's renomination. He was deeply aware of the dark mood of America and he fully shared the sense of many that the country, somehow, had lost its way in the four and one half years since the bright promise of his brother's leadership had been snuffed out by an assassin's bullet in Dallas. And when all was said and done he too placed much of the blame at the feet of his brother's successor—though he went out of his way to avoid personality clashes with Johnson, to a degree few outside his inner circle ever realized.

It was inevitable, in the light of the history of unhappy personal relations between the two men, that much of Kennedy's issue-oriented criticism was interpreted almost entirely in terms of the Kennedy-Johnson feud. That fact caused Kennedy great consternation throughout 1966 and 1967, and no more so than in the fall of 1967, when he began seriously to weigh the possibility of challenging the renomination of an incumbent President. There was no doubt that he wanted to depose Johnson, nor could there be any doubt that he considered the President the usurper of his brother's New Frontier. But Kennedy needed, for himself, to move as his own man, for his own good reasons unrelated to the bitterness of the past. He was convinced—or at least had rationalized—that what was driving him chiefly was not personal pique toward Johnson, but a broader, impersonal certainty that Johnson's policies at home and abroad were tearing the country apart. The question was: could he run against Johnson on the issues, without having the whole effort dismissed as a personal vendetta, thus splitting the Democratic Party and the country wide open, delivering the

White House to a Republican and, in the process, destroying his own political future?

The feud, of course, was very real, despite periodic efforts by both men to deny its existence. In background, education, modus operandi, circle of friends, heroes and foes, the two were utterly different men. Kennedy, as everyone in public life knew, hadn't wanted Johnson as John F. Kennedy's running mate in 1960, and Johnson had rejected Robert Kennedy as his own Vice Presidential candidate in 1964. Moreover, there always was the bitter irony of this crude Texan—flamboyant, earthy, insensitive to subtlety—assuming the Presidency that had been so elegantly carried by Robert Kennedy's dashing, polished, sophisticated brother. For the public, no one seemed to have a greater, more justifiable dislike for the man who gained the White House as a result of the Kennedy family tragedy than the younger brother who once had been the New Frontier's second most influential figure.

Actually, both Kennedy and Johnson often expressed regret to associates that there had to be ill will between them. Still, their relationship, purely in terms of governmental responsibilities, was a sore point to each when the other had the upper hand. Johnson resented the fact that Kennedy during his brother's Administration had continued to consider him to be outside the President's close inner group; Robert Kennedy as holdover Attorney General under Johnson and later as a Senator chafed at the short shrift he received. When Kennedy returned from fact-finding trips to Latin America and other places in placid times he never was asked to tell Johnson or the State Department what he had learned. But when he went to South Africa as a Senator in 1966 and won wide acclaim for his outspoken opposition to apartheid there—at a time Johnson was being mightily embarrassed by Senate Foreign Relations Committee hearings on Vietnam—the summons finally came. Stopping at his New York apartment on his return, Kennedy got a phone call from Walt W. Rostow, Johnson's chief national security adviser, asking him to come to the White House that night. It was the first such invitation to Kennedy since he had left the Johnson Administration, and he was bitter. "That s.o.b.," he told a reporter when he hung up the phone. "All the other times I came back with something to tell, and he never asked me." Kennedy deliberately waited until the next day to accede to Johnson's request.

Yet the impact of the feud on the public was a source of continuing concern to Kennedy. It was not because he found fault with the premise that he did indeed dislike Johnson, but because he saw the feud explanation as a great oversimplification for his dissent, robbing him of a political identity of his own. "He couldn't look at Lyndon Johnson without the eyes of the country trained on him," a Kennedy intimate recalled later. "And he felt it more acutely because it had some basis in truth." Nor was Kennedy's concern without personal political motivation. He was fully aware that one of his own major political problems was the public impression that he was "ruthless"—that he acted out of personal vindictiveness or righteousness to get his way.

Unlike other politicians who deal with the electorate from the start of their careers, Robert Kennedy had been essentially a private man for most of his sixteen years in government. His only real constituency was his brother John, and cast in the role of a behind-the-scenes lieutenant, his function was not to be popular, but to get things done. After 1963, when he decided he must become a public man with his own constituency in order to pursue his aims, all that changed. The manner in which he did things, and how he dealt with people, became a matter of great public interest. Robert Kennedy soon learned this fact, and became painfully aware that his early brashness had left a scar on his public image. He labored to remove it, tried to joke it away and was intimidated by it. It became a key element in his feud with Johnson; the nature of that relationship, and of Johnson's ascendancy to the White House, always kept that scar in public view. Any move against Johnson and toward the Presidency itself would be interpreted by many as entirely personal and self-serving.

And beyond the political damage this interpretation could do to his chances of success, there were other equally deep-seated apprehensions: that the fiber of his criticism would be dismissed without consideration of its merits, that the substance would be lost in the reflex reaction of attributing it out of hand to the feud; that a Kennedy candidacy might harden Johnson's position on the war, or even lead him to escalate the fighting and the stakes; that Kennedy would sacrifice his seemingly inevitable political rise to power for nothing.

There probably was not a major policy statement Kennedy made in the Senate from 1965 through 1967 that was not carefully reviewed by him or his chief aides to focus criticism on the actions or policies of Johnson, or on a problem, rather than on the personality in the White House. But the public and the press were conditioned by then to gauge the current temperature of the feud through Kennedy's speeches. Because he was attacking or at least express-ing reservations about a spectrum of Johnson policies, it was certain that each new statement would be met with a rash of stories on the latest chapter of the Kennedy-Johnson personal breach. Perhaps Kennedy was naive to expect any other reaction, but it constantly irked him.

Kennedy's first major break with the Johnson Administration over the Viet-nam War came in his February, 1966, statement that to admit the NLF "to a share of power and responsibility" in the Saigon government was "at the heart of the hope for a negotiated settlement." Kennedy had labored and agonized long over the speech and had talked to many friends in the Administration about it, trying to keep it issue-oriented and on target. But the statement was viewed immediately in the press as a soft-line escape scheme and the deepest evidence yet of the Kennedy-Johnson feud. He spent the next several days seek-ing to counteract those impressions, even to the point of conferring with John-son's press secretary and his own friend, Bill D. Moyers, about ways to narrow the gap that had been opened by the statement.

A year later, Kennedy returned from Paris after talks with the French for-eign office and was called on the White House carpet by Johnson for suppos-edly indicating he had picked up a Vietnam peace feeler there. He denied it in a stormy forty-five-minute meeting that also added much fuel—justifiably—to the feud talk. The session reinforced Kennedy's concern about his ability to ef-fectively oppose the war in the face of his relationship with Johnson—and his impression that any time he criticized the war, it only hardened the President's policies. The Senator took the public posture that he supported Johnson for reelection, a close confidant recalled later, "because he felt it liberated him to criticize the war—that it took his criticism out of the bounds of personality. It may have been inconsistent, but that was how he felt."

A week after that early February, 1967, White House meeting, Kennedy gave a major speech at a University of Chicago seminar on China, urging "a search for a new China policy." As he prepared the speech, the low ebb of his always rocky relationship with Johnson was much in his mind—and in the conscious-ness of the press and public. On the late afternoon of the day of delivery, the advance text was provided in Washington to Andrew Glass, then of the Washington *Post,* and me, the only two reporters who would be accompanying him to Chicago. There was no general distribution of the text; it was given to us because a late and hurried departure was anticipated and the lead time enabled us to write early stories before leaving. As we prepared our accounts, Kennedy and aides reviewed the speech text in his Senate office. It was decided to insert a phrase specifically absolving any one President for American policy failures toward Communist China—an obvious effort to keep the focus on the substance of the criticism and off the feud. Glass and I were advised of the change by Frank Mankiewicz, Kennedy's press secretary.

Riding to Dulles International Airport in Kennedy's car shortly afterward for the evening flight to Chicago, the Senator asked Glass what he had found in the speech to write about. My colleague told him that Kennedy had attacked Johnson's China policy—but had inserted a last-minute softener in his advance text. It was true enough about Kennedy adding the softener, but the first version of the advance text had been given to us as a courtesy; only the final version, incorporating the disclaimer, was officially released to other newsmen. Thus, the press generally had no reason to know or to say that the phrase in question had been "inserted" to soften the blow. Kennedy, when he heard what Glass had written, groaned and began to argue heatedly with him.

"Is that all you could find in there?" Kennedy asked, pained. "Wasn't there anything in there about China worth writing about?"

The *Post* reporter said he thought there had been a general release of the text, and that the softener was indeed a change in the released version. But when Kennedy continued to berate him, he agreed to telephone a correction to his office from Dulles before the plane left. On arrival in Chicago, Charles Daley, a Kennedy aide there, had the original story off the wire of the Washington Post-Los Angeles Times Service. The Senator resumed his lament to Glass as we walked into the University of Chicago hall for the speech. "It will go out all over the country as only another attack on Johnson," Kennedy wailed. Afterward, in a question-and-answer period for the students, Kennedy took pains to speak well of Johnson. He declined to dispute the President's standing contention that he had inherited the Vietnam War; it would be "very unfair," Kennedy said, to saddle Johnson with blame for it. The effort to stifle the feud aspect of his China criticism was obvious, but as always the attempt was unsuccessful, as the next day's newspaper stories in major cities confirmed.

The press' focus on the feud, in Kennedy's view, inhibited his own effectiveness as a critic not only on Vietnam, but also on major domestic issues about which he had legitimate cases to make. In October, 1966, he went to Berkeley for a civil rights speech on which he and his staff had devoted much time and thought. The event occurred on the eve of talks between Johnson and then South Vietnamese Premier Nguyen Cao Ky in Manila. Kennedy was asked about Ky in the question-and-answer period. He said he didn't think the people of South Vietnam really wanted Ky as their leader. Some reporters subsequently wrote that Kennedy was undercutting Johnson on the eve of a critical conference—although the Senator earlier had told a questioner flatly he would not "dissociate" himself from Johnson because of Vietnam. Kennedy, concerned that the feud would be magnified and would obscure a civil rights speech whose message he considered to be important, walked into the press room and tried to dissuade David Broder, the highly respected chief political writer of the Washington *Post,* from featuring the exchange about Ky. Broder refused. Later, a Kennedy aide called his editor to block publication. He failed; but the story was moved from page one in first editions to an inside page later on.

In understanding Kennedy's agonizing in the fall of 1967, when the role as the candidate of dissent was his for the asking, this obsession that his own reasoned disaffection was being dismissed as a personal hate against Johnson must be given full weight—as self-serving as it may have been as a rationalization for inaction.

That a Kennedy would need such a rationalization, or justification, for holding back is not hard to appreciate. The whole fiber of the family and of its men, the lesson that deep and repeated tragedy had taught its members over the years, was that there might be no tomorrow. This awareness thrust John F. Kennedy into the race for the Democratic Vice Presidential nomination in 1956, at the age of thirty-nine, and into a Presidential bid at the age of forty-two. Older and supposedly wiser counsels who preached a more patient political course were brushed aside, as they were by Teddy Kennedy when he ran for the Senate in 1962 at the age of thirty—the constitutional minimum—and by Robert Kennedy in 1964, when he ignored tradition and became a Senatorial candidate from a state in which he had not been living. The Kennedy family dictum was to grab opportunity when it came along, and to run with it. Joseph P. Kennedy, Sr., its patriarch, had instilled this spirit into his children; the series of deaths and near-deaths that visited the clan reinforced it for the survivors.

In Robert Kennedy, as in the others, there was a strong sense of fatalism, along with a seeming compulsion to take maximum advantage of the time he had. It showed both in the range of his activities and in the intensity and daring with which he pursued them, be it his "carpetbagger" bid for the Senate in New York, climbing newly named Mount Kennedy in Canada, shooting the rapids in the Rockies or keeping five or six special projects going on the side as he pursued his main job as Attorney General or as Senator.

He spoke occasionally about his fatalism; not voluntarily, but sometimes when he was asked about what he would do in the future. In October, 1966, at a gymnasium rally at Sacramento City College on a bright Sunday morning, someone urged him to give "some assurance that you will run for President in 1972." Kennedy paused, then said: "Oh . . . well, I . . . just quite frankly don't know what the future brings. I think that one cannot plan that far in advance. I'm going to continue, as long as I'm around on this globe—I'm going to continue in public life in some way . . . I don't know when that man way up there is going to take me, so I can't . . . that's not a very satisfactory answer, but it's the best I can do."

As fatalistic as Robert Kennedy was, he was not compelled to throw his bright political future away for nothing. However, there could be little doubt that it would be politically safer to stay out of the 1968 Presidential race and wait until 1972—although there were those who warned him that his great public appeal, in part at least a carryover of the feeling for his late brother, would not last forever. The logic of caution obviously was exceedingly prominent in the overall equation. But so had it been in other key political decisions that had faced the Kennedys, and in all the big ones they had cast caution aside and risked all on the opportunity and the challenge of the moment. Indeed, it was this family characteristic that made so many Americans irritated, even frightened, by the Kennedys' pursuit of national power. In a special foreword to a young readers' memorial edition of John F. Kennedy's *Profiles in Courage* written less than a month after the President's assassination in 1963, Robert Kennedy had observed that his brother was fond of quoting Dante that "the hottest places in Hell are reserved for those who, in a time of great moral crisis, maintain their neutrality." In the fall of 1967, Robert Kennedy was feeling the fires with increasingly embarrassing intensity and self-doubt.

Actually the Senator had been weighing the subject of a 1968 candidacy all through 1967, though not with any real expectation that he would challenge Johnson. Yet the polls showed him at times running stronger than the incumbent President, and the growing dissent against the war within the Democratic Party was looking for a leader, a candidate. A "Citizens for Kennedy-Fulbright" group was started by a bumbling political amateur, Dr. Martin Shepard of New York, in the spring of 1967. It made plans to enter a slate in the New Hampshire primary and held a conference in Pittsburgh in late September. But it was a bush-league operation and Kennedy wanted no part of it. Also in the spring, a young New York lawyer named Allard K. Lowenstein, a knowledgeable Kennedy ally in the byzantine entanglements of New York Democratic reform politics and an all-out critic of the Vietnam War, spoke with Kennedy repeatedly about the need to dump Johnson. Knowing Kennedy's stormy relationship with the President, Lowenstein assumed that Kennedy would not try it himself, but he discussed the prospects with him.

On a flight to California in August to attend a dinner for State Assembly Speaker Jesse Unruh, Lowenstein laid out to Kennedy how Johnson could be beaten in the primaries. A drive to muster a resounding "No" vote against the President in Wisconsin was one approach; the running of a peace slate in California was another. Kennedy was sympathetic, but thought it quixotic. "He took it as seriously as the idea of a priest in Bogotá deposing the Pope," Lowenstein recalled later. Kennedy did, however, ask Lowenstein whom he was going to approach to run. "Gavin," the New York lawyer said, referring to retired Army General James M. Gavin, proponent of the "enclave" theory for maintaining an American presence in Vietnam while negotiating peace. "If you can get him," Kennedy said, "you're really in the ball game."

Lowenstein set out to find his candidate. He went to Boston and tried Gavin, who informed him he was a Republican. Next, he went to former Ambassador John Kenneth Galbraith in New York. "He thought it was an extremely good idea," Lowenstein said afterward, but Galbraith was born in Canada of Canadian parents and couldn't overcome the constitutional barrier. Lowenstein consulted with other Vietnam critics, including Richard N. Goodwin, the former John Kennedy and Johnson speechwriter, and they all suggested two other possibilities: Senators George S. McGovern of South Dakota and Eugene J. McCarthy of Minnesota. Lowenstein already had them on his shopping list and by happenstance got to McGovern first. He saw the Senator several times and McGovern was sympathetic. But although he never gave Lowenstein a flat no, his mind was on reelection to the Senate in 1968. McGovern also mentioned McCarthy, suggesting he was better known, and urged Lowenstein to see him. McCarthy's first response, according to Lowenstein, was, "I think Bobby should do it." But he too was sympathetic, and it went on like that for several weeks—an ad hoc search for a dump-Johnson candidate, marked by exploratory conversations with several prominent Vietnam critics at any one time.

On October 15, a Conference of Concerned Democrats was held in Chicago to give impetus to the effort. Senators Frank Church of Idaho, Vance Hartke of Indiana, McGovern and McCarthy were recommended as possible anti-Johnson candidates. Lowenstein, a co-chairman of the conference, kept Kennedy posted on the antiwar organizations growing in several states, cajoling him all the while about "time running out." Two or three times he pressed Kennedy seriously about the country's need for a candidate to take on Johnson, but he never directly asked Kennedy to do it. "I didn't have to spell it out," he said later. "He knew I wanted him, but if he was going to go, he didn't need me."

McCarthy, though, was a virtual unknown, and he did need somebody. Lowenstein had more talks with him; the first inkling that he had found his candidate was when McCarthy asked him, "How do you think we would do it?" From then on, the talks changed from whether to run to the tactics of running. In the end, Lowenstein didn't persuade McCarthy to be the anti-Johnson candidate as much as he made the decision possible by providing the embryo of an organization with which McCarthy could make a start. To McGovern on the Senate floor one day, McCarthy said, "You know those people you sent over to me? I may just do that."

For Kennedy, though, it was not outside pressure, but inner pressures building up from the war and urban conditions at home that forced him to think more seriously about a Presidential bid as the fall wore on. And it was inevitable that Robert Kennedy, trapped in his deep concern for the state of the nation, the risks to his political future and his role as an effective critic of the Johnson Administration, would turn first to members of his family for counsel. To run or not to run was a personal decision, but as always it was also a family decision. For the women—Ethel, his wife, and sisters Eunice Shriver, Jean Smith and Pat Lawford—the reaction was mostly visceral and emotional. They wanted him to regain the lost castle, and said so. But the one on whom he relied most heavily now, brother Teddy, was not only a Kennedy but a Democratic politician, and the pragmatism of the latter restrained the political acquisitiveness of the former. He too was aware of the public impact of the Kennedy-Johnson feud on a meaningful discussion of the war and other issues; aware of the political dangers to his brother—and his party—of a fight for the nomination that would be taken by so many, as Robert Kennedy himself put it, "as a personal vendetta" against the duly elected President.

Just as Robert Kennedy in 1960 had been the one man on whom his older brother had leaned most in his Presidential campaign, he in turn looked to Teddy, six years his junior, in this first serious facing-up to possible Presidential candidacy in 1968. But unlike the 1960 parallel, Teddy Kennedy at this juncture was a national political figure of stature and influence in his own right, not only attuned to the pressures for party regularity and unity but subject directly to them. He agreed with his brother's criticism of Lyndon Johnson's Vietnam policy. But he also was among the strongest believers that Robert Kennedy could not run against Johnson without the effort—and the issues—being lost in the public furor over "the feud," and over the "ruthlessness" of a boyish freshman Senator reaching out impatiently to replace an incumbent President of his own party.

Besides, the younger brother was much more an organization Democrat than Robert Kennedy ever had been, particularly in the Senate. Robert from the first was a loner, disregarding Senate tradition about freshmen being seen but not heard, and chafing at the slow pace of the legislative process. Teddy by contrast played the Senate game, won the admiration and affection of the older men to whom the wheel-spinning, the empty gestures and the time-wasting procedures were cherished trademarks of "the club." This closeness to party regularity, and the consequent awareness of its potency against anyone who dared challenge it, set Ted Kennedy firmly against a 1968 Presidential try by his brother in these first discussions. It was an opposition that was to persevere almost to the end of the months of soul-searching.

Beyond the family, there were three other broad groups of confidants involved in the run-or-not-run discussions, though they never met with Robert Kennedy as formal groups, only as individual friends and associates in politics. In each there were those who wanted him to get in and those who wanted him to stay out. The first group of the three was composed chiefly of 1960 campaign holdovers who, like Teddy, also wanted to see Robert reinstate his late brother's New Frontier sometime in the future. But some of these men, also like Teddy, now had a stake of their own in the Democratic Party; most agreed at the outset that only political disaster lay ahead in a 1968 candidacy against an incumbent Democratic President. In this group were Stephen E. Smith, the Kennedy brother-in-law and a chief 1960 money man and strategist: Theodore C. Sorensen, John F. Kennedy's White House counsel and chief speechwriter; Pierre Salinger, the former Presidential press secretary; Kenneth O'Donnell, President Kennedy's appointments secretary; Dick Goodwin; Frederick G. Dutton, former White House adviser and State Department liaison with Congress; Arthur M. Schlesinger, Jr., the historian and former Presidential speechwriter; and other lesser lights of the New Frontier. Of these, only Schlesinger and Goodwin were very early advocates of a 1968 candidacy.

In the second group were Kennedy's old Justice Department associates, most of whom favored his candidacy, especially in the later stages of the deliberations: Burke Marshall, former head of the Civil Rights Division; Edwin O. Guthman, Kennedy's press secretary at Justice; John Seigenthaler; John Nolan; John Douglas, son of former Senator Paul Douglas; William vanden Heuvel of New York and others.

In the third group were Kennedy's Senate staff members, most of whom were gung-ho for him to run: Adam Walinsky, his young, energetic, aggressive, sometimes impatient speechwriter; Peter Edelman, his mild-mannered, quietly impressive legislative researcher; Frank Mankiewicz, Kennedy's witty, efficient press secretary; Jeff Greenfield, another young speechwriter and researcher; Michael Schwartz and Lewis Kaden, legislative and issues research-ers; and others. Joseph Dolan, Kennedy's administrative assistant in the Senate, really qualified as a member of all three groups, having also functioned as a John Kennedy campaign operative in the Far West in 1960 and as a Deputy Assistant Attorney General under Robert Kennedy at Justice.

In the earliest soundings, Kennedy turned first to the old 1960 hands, his Justice buddies next and, generally speaking, his Senate staff much later. He had been through the fire of Presidential campaigning with the members of the first group, so it was only natural that he would seek their advice first; he had been through sticky domestic, racial and personal—but not necessarily political—crises with those of the second; he had spent the shortest time with those of the third, and besides, he knew how they felt. Walinsky, stumping the country with Kennedy in 1966 for Congressional candidates, got one look at the wild outpouring of enthusiasm for his young boss, climbed aboard the plane and, disregarding the fact it was not a Presidential election year, pro-claimed: "The hell with 1968! Let's go NOW!" Kennedy knew if he were to run he would have to have the others—the old pros—with him. And wanting to make the race but not seeing how he could put it together, he asked those who, if anybody, could.

Nolan, now a Washington lawyer, reflected later on Kennedy's dilemma. "In his gut, he wanted to go," he said. "He felt it was the right thing to do, but his judgment told him his chances of success were so slim, it wasn't the right decision. From the neck down he wanted to go; from the neck up it just didn't make sense to him."

It was the Kennedy style—both John's and Robert's—to avoid "critical" decision-making meetings of set committees. Group conversations on Robert Kennedy's possible candidacy did take place in his presence in September and early October, but not at his calling nor with much participation by him. One night at Hickory Hill, Kennedy's home in McLean, Virginia, Lowenstein, Goodwin and Jack Newfield, a New York writer and spokesman for the New Left, entered into a debate of sorts with Schlesinger. The first three stated the case for his 1968 candidacy, Schlesinger at that time was against it. He was ar-guing instead for a campaign to impose a Vietnam peace plank on the Demo-cratic convention in 1968. Kennedy just listened impassively, but finally broke in. "Arthur," he said, "when was the last time you heard of people rallying to a plank?"

When shortly afterward Kennedy set about picking the brains of his closest friends and associates on what he should do, he did it in scores of individual conversations—in his Senate office, at Hickory Hill, in his New York apartment on United Nations Plaza, in airplanes, on the telephone. Seldom did one contact know the other had been called, yet the consensus of these soundings was solidly against a 1968 race. The men approached, like Kennedy himself, were winners in whatever they chose to do, in politics or some other field. They were not, nor was he, interested in windmill tilting, and the odds against dumping an incumbent President, on a platform anchored to a still-unpopular dissent from Johnson's war policies, seemed to most of them impossibly long.

Still, the Kennedy team wanted to retake the castle sometime, and some old members, including Salinger, finally thought in mid-October that it was time to meet and explore the pros and cons together, just to know where they stood. "God only knows what's going to happen," Salinger told Kennedy one night during dinner at Hickory Hill. "But we ought to be realistic." Kennedy was lukewarm to the idea, but finally agreed to let Salinger go ahead, on the quiet.

The first meeting of any sort that deserved the description took place on a Sunday in late October in New York, in Salinger's suite at the Regency Hotel. Most of the old hands and a few others attended. They included, of course, Ted Kennedy and Steve Smith; Pierre, Fred Dutton, Kenny O'Donnell, Goodwin, Ted Sorensen, Schlesinger, Chuck Daley—all members of the ingroup. The presence there of a few baffled the others. One of these was Ivan Nestingen, a rather obscure Undersecretary of Health, Education and Welfare under John Kennedy, who figured prominently in the 1960 Wisconsin primary campaign. O'Donnell, knowing Nestingen was against the war and understood the Presidential primary route, brought him along. John Burns, the New York State Democratic chairman and a more recent Robert Kennedy ally, came later.

Significantly, the Young Turks of the Senator's office—Walinsky, Greenfield and the like—were not asked. "He knew how he felt on the issues," Goodwin recalled later. "He was looking for political judgment." There was some talk about the wisdom of Kennedy himself attending, but it was ruled out, although the Senator in the end wanted to go and asked his brother and others whether he should. The verdict was that he had better not; his attendance, if it leaked out, certainly would make the meeting appear to be a major conspiracy to depose Johnson. O'Donnell specifically told Salinger he would not attend if Kennedy did. Also, if he were there, the participants might be inhibited in what they chose to say. So he stayed away, and afterward Teddy and Steve Smith reported the outcome to him.

Predictably, the consensus, with the exception of "outsider" Nestingen—who wanted him to run full blast—was against Kennedy's running. But among the others, the same reasons he himself had often voiced were echoed: that it would look like a personal vendetta; that it wouldn't end the war and might only stiffen Johnson's back and lead to escalation; that Kennedy would look foolish and "ruthless"; that it couldn't be pulled off. And for the pros, that last was enough. According to Salinger, a private poll was conducted in New Hampshire, and the result was sobering: Johnson 57 percent, Kennedy 27 percent. Others denied knowledge of such a poll, but they didn't need it to be persuaded against Kennedy's running.

Kennedy knew by this time that Senator McCarthy probably was going to agree to be the candidate of the Lowenstein dissidents. Salinger, back in California, went to a party for McCarthy given by Gene Wyman, the Democratic state chairman. McCarthy spoke and it became clear at once to Pierre that he was listening to a man getting ready to run for President. Afterward, he asked him, "Gene, what day are you announcing?"

"What day would you suggest?" McCarthy replied.

The next morning Salinger phoned Kennedy. "Guess what," he said. "We've got a Presidential candidate."

"Who?"

"Gene McCarthy."

"You're kidding," Kennedy said.

It could not, of course, have come as that great a surprise, because Lowenstein had been keeping Kennedy advised. At any rate, shortly before announcing his candidacy, McCarthy contacted Kennedy and told him of his plans. Some McCarthy aides who were privy to reports of the conversation said later McCarthy had indicated he would not run if Kennedy would, but that Kennedy had said no. According to McCarthy himself, he never directly asked Kennedy to run.

When McCarthy finally did announce his candidacy on November 30, he noted however that he had waited "a decent period of time for others to indicate" they would make the test. "I would have been glad . . . to have had him [Kennedy] move early," McCarthy told a questioner. "I think if he had, there would have been no need for me to do anything."

McCarthy even seemed to be suggesting that he wouldn't be too upset if Kennedy came in later on and took over. Asked if he thought Kennedy might move in later if McCarthy made "a significant showing," the Senator replied: "Well, I don't know. He might. It would certainly be nothing illegal or contrary to American politics if he or someone else were to take advantage of whatever I might do or what might happen in consequence of what I'm doing. . . . There's no commitment from him to stand aside all the way and it would certainly be in order for him and only proper, it seems to me—within the rules of American politics—for him to make that kind of move." Well, would he step aside if Kennedy came in? "That's projecting things a long way ahead," McCarthy said. "I don't see that as a problem right now. Again, I don't know whether it would be a question of stepping aside. It might be more—it may be less voluntary than that. But I don't see that as a great disaster, let me tell you, if it should happen that way."

In light of this answer, Kennedy had good reason to believe that if he finally decided to run, disposing of McCarthy wouldn't be too difficult. (McCarthy's pliability on the point didn't sit too well with some of his young supporters, and he soon moved to assuage them. In a by-lined article in *Look* magazine dated February 6, 1968, but written several weeks earlier, McCarthy said that if Kennedy planned to let him fight the primaries and then move in, "he will have a fight on his hands to see who has the most strength. I will not step aside voluntarily."

But McCarthy's candidacy in itself applied more heat to Kennedy. Not only did it provide an embarrassing contrast to his own caution against running, but it made his position of support for Johnson's reelection all the more inconsistent. On ABC's *Issues and Answers* television panel on January 7, McCarthy said that "so far as Senators who have supported the position I take who were not running [for reelection], I think there is a little bit more burden upon them to take a stand [against Johnson], and the excuses made by some that I am not going to win, I don't think that is the kind of excuse that really stands up very well." Kennedy finally adjusted to the new realities by saying he would support the nominee of the Democratic Party—thus leaving himself on more defensible ground with the McCarthy camp, if not with the Administration.

All this time, while Kennedy seemed to have dismissed serious thought of running, he was being pressured by his friends in the antiwar community and by the quagmire of Vietnam itself. He had been thinking about a trip to Eastern Europe in December, and he asked John Nolan, who had advanced all his overseas trips when he was Attorney General, to plan that one for him. Nolan looked over the situation and decided that Eastern Europe in December was the wrong place at the wrong time for a politician who might be running for President in 1968. On Nolan's advice, and because a drive for a new state constitution in New York recommended his presence, Kennedy decided not to go.

Nolan, who had made arrangements to leave his practice long enough to advance the Robert Kennedy trip, was asked instead by Ted Kennedy to go to Vietnam to arrange his scheduled late-December exploration of refugee and civilian casualty problems. Nolan went, in early December, and came back convinced the war and American policy were all wrong. He found that the refugee program was complete chaos and that civilian casualties were four or five times greater than American authorities were admitting. Nolan relayed his feelings to Robert Kennedy and suggested he go to Vietnam himself. He would learn how bad things were, Nolan said, and the trip also would give what Kennedy said about the war an added ring of authenticity.

Nolan wasn't ready to urge Kennedy to run, but he argued that the political situation was fluid and that Kennedy ought to prepare himself as best he could if later circumstances pulled him into the race—a race in which Vietnam obviously would be the key issue. But Robert Kennedy held off.

About two weeks after McCarthy announced, the Kennedy insiders met again in New York, this time with Robert Kennedy present. The meeting actually was in two shifts, a late Sunday morning brunch session involving most of those who had been at the October meeting, and a second session for these and a larger group in the early afternoon. The October meeting had found its way into the Evans and Novak column, and Kennedy hoped somehow by this format to avoid a repetition. The first meeting—which was held without the general knowledge of those who were invited only to the second—took place in Bill vanden Heuvel's apartment on Central Park West and was called by him or Steve Smith, with Kennedy's acquiescence. How vanden Heuvel got so deeply into the inner circle always was a mystery. He was regarded by many of the others as a political lightweight, but he was part of the Kennedy social set and he often functioned in the political ranks. John Kennedy always had kept the two groups apart; Robert Kennedy let them mix as they would, and in this permissive climate, vanden Heuvel became a blur between the two. Others attending the insiders' meeting were Ted Kennedy, Salinger, Sorensen, Schlesinger, Dolan, Dutton and Goodwin.

Schlesinger and Goodwin made the case for Kennedy's entry into the race; Salinger said later he was for it too, but others said he was on the fence. Most still were against the idea—and especially Sorensen. Vanden Heuvel favored trying to force Kennedy's nomination as Johnson's running mate, but Kennedy dismissed it. He would lose his constituency for 1972 for nothing, he said. To at least one participant, Schlesinger was the most persuasive on running for the Presidency. "Arthur was all-out for going," he recalled afterward. "He argued that maybe it couldn't be done as a practical matter, but the war was terrible and Johnson was no good, so Kennedy had to put personal considerations aside. He said historic things were happening, and Bob owed it to the kids to get in."

Kennedy himself was itching to go. When somebody said a 1968 candidacy would sacrifice his political future, he snapped: "My future is not the issue. The issue is whether the country can survive four more years of Lyndon Johnson. If by declaring myself a candidate I could end this war any sooner, I'd feel an obligation to run." But he was aware of the practical pitfalls. "I'm against the war," he said at another point, "but if I get in I want to have influence. What will Johnson's reaction be? Escalate the bombing? And everybody will believe I'm waging a personal vendetta." It always came to that. "He wanted to run like hell," another of the participants said later. "But he wasn't interested in abstract crusades unless they were effective."

McCarthy's impact also was discussed at length. Kennedy indicated he thought McCarthy didn't have the political muscle to wage the fight against Johnson, and that if anything effective was to be done, he himself would have to get in. Although Kennedy didn't say so, he seemed confident he could get McCarthy out at the proper time. Only recently, Kennedy told a close aide, McCarthy had come up to him in the Senate and advised him he wanted only one term in the White House, and would step aside for Kennedy in 1972!

In the second meeting, in the office of a shipping firm near Madison Avenue run by one of vanden Heuvel's clients, the Senate staff types—Walinsky; Edelman; Milt Gwirtzman, a lawyer-researcher; Dave Burke, Ted Kennedy's administrative assistant—and others including New York politicans John Burns and Jack English had their say. Walinsky and other Senate staffers wanted their boss to run, but the consensus still was negative, and few wanted to push Kennedy over the cliff. "Everybody was a Junior Abe Fortas," one of the participants later noted. "Nobody wanted to say what to do; everybody was just laying out the alternatives to be analyzed. Bob, as always on major matters, was deliberate, wary, walking around all sides of the question." And from most angles, the view the potential candidate got spelled futility and political disaster. So he remained out.

Around the same time still another insiders' meeting was held at Hickory Hill, attended by Robert and Ethel Kennedy, Jesse Unruh, Kenneth O'Donnell, Fred Dutton, Burke Marshall, and a few others. Ethel asked each of them what her husband should do. O'Donnell's answer was typical of all the responses: "I'm not advocating it, but if he announces at twelve noon, I'm with him at twelve one." O'Donnell was an early critic of the Vietnam War, but he was also a hard-nosed politician. He put the odds against Kennedy at five to one; Unruh said they were three to one, but he would even resign as California Speaker to help him. These were sobering judgments.

About this time Kennedy also talked to O'Donnell about the possibility of entering the New Hampshire primary as a kind of test run. O'Donnell opposed it on the grounds that it would look like an appeal to the past and could too easily be written off as a backyard victory for the New England–reared brother of a New England President. If Kennedy was going to run, O'Donnell told him, the only way was to jump in with both feet in every primary. "When you go,"

O'Donnell said later, "you don't play around. You take your big gun out of the barn and fire it." This Kennedy was not ready to do.

In early January, Ted Kennedy as head of a Senate subcommittee on refugee problems made the Vietnam investigative trip that Nolan and others had advanced. He returned with the same reaction Nolan had transmitted to Robert Kennedy—that the refugee program was a colossal fiasco, reflecting the state of the war itself.

On a Sunday in mid-January, Ted dropped by Hickory Hill and spoke at length and with great feeling about all that he had seen wrong about the American role in Vietnam—especially the corruption and inefficiency among refugee officials and the police. All this only added to Robert Kennedy's conviction that he was right on Vietnam and Lyndon Johnson was wrong. But it still seemed sheer folly to challenge him for the party nomination.

Also in Kennedy's thinking now was a conversation he had had in early January with Jesse Unruh in California, where Kennedy had gone with Peter Edelman for some Senate committee hearings on Indian problems. Unruh, then considering whether to challenge Republican Senator Thomas Kuchel, was commissioning a poll to determine his chances. He told Kennedy he would include questions about Kennedy's standing in the state and urged him to delay any firm decision until the results were in, in about four weeks. In any run for the Presidential nomination, the primary in California, which had 174 delegate votes, obviously would be critical. Together with the 190 convention votes from Kennedy's own state of New York, the California delegation would give him a powerful base from which to make his bid.

All this time, attitudes within the inner group were hardening against the war, but none of the insiders could see how he could be successful. The consensus answer remained "No." Kennedy could do as much as was possible, most of them believed, by continuing to speak out as a Senator who was loyal to his party but against the existing policy on Vietnam.

It clearly was not an emotionally satisfying judgment for Kennedy, but his head would not let him fight it. So he went on as usual, persuaded as those he most trusted were, that challenging Johnson would be an exercise in supreme futility. At home, the decision did not go down well. Ethel Kennedy prodded him to run. She was as much against the war as he was, and she felt he really wanted to make a try. So she needled him to get in where the action was. Friends suspected Ethel really was reading her husband's own desires and doing what she could to help him act on them.

Actually, she had only to read the postscript of his new book, *To Seek a Newer World*. In it he challenged youth to effect change, but warned of four dangers: futility ("the belief that there is nothing one man or one woman can do against the enormous army of world's ills"); expediency ("those who say that hopes and beliefs must bend before immediate necessities"); timidity ("few men are willing to brave the disapproval of their fellows, the censure of their colleagues, the wrath of their society"); comfort ("the temptation to follow the easy and familiar paths of personal ambition and financial success so grandly spread before those who enjoy the privilege of education"). Few would have accused Kennedy then of yielding to the comfortable life. But the other pitfalls he identified in the book's peroration were stinging self-indictments of his caution.

The pressures from the Young Turks were plaguing him, too. Talking in his Capitol Hill office with O'Donnell one day, Kennedy asked him, "Do you think I can win?"

"No," O'Donnell said flatly.

"Well," Kennedy replied, "will you stay and talk to Edelman and Walinsky? I like these fellows, but they're out in left field on this. Please explain to them. Tell them the facts of life."

O'Donnell did so, and later he recalled: "They said, 'So what?' And I privately agreed with them. It was worth running and losing for. But what right did I have to tell a guy to run when I was saying he couldn't win? I had no right to say to a guy, 'Run and lose.' "

It was now late January and the possibility that Kennedy might yet enter the race had not been completely laid to rest in public. A long-scheduled background breakfast with a group of Washington political writers was approaching, and Kennedy and Mankiewicz knew the question of his 1968 plans was certain to be asked. It seemed as good a time as any to clear the air.

Driving to the breakfast at the National Press Club on January 30, Kennedy and Mankiewicz weighed what the Senator should say. At the backgrounder, it was agreed that the discussion would proceed off the record and then the participants would decide at the end what would be placed on the record, for direct attribution to Kennedy. It was under these ground rules that the question of his plans did come up, early in the meeting, when he was asked whether there were any circumstances under which he would be a candidate against Johnson. His negative response, and his comment that only "an act of God" would keep the President from seeking reelection, convinced most of the reporters present that Kennedy really was ruling himself out.

The talk that he was considering it at all, Kennedy said, was McCarthy's fault. "I think McCarthy started the speculation when he said people who feel strongly [about the war] on principle should speak out," he charged. After that, Kennedy said, more and more people started coming to him about it. "My feeling is that after it was all over and gone through, this [his candidacy] would be very damaging in my trying to speak about any of these matters. I have to talk in practicalities. If I ran, a lot of states would be split down the middle. I not only would take the risk of weakening my opportunity to talk and have an effect on the issues, but I could bring down a good many other Democrats as well."

In all this comment, Kennedy made it clear he didn't think much of McCarthy or of his chances. "I don't think his campaign is going very far or very well," he said, speaking six weeks in advance of the New Hampshire primary. "His campaign so far has been very helpful to President Johnson. . . . It could have been and still is to some extent an outlet to frustration about the war. The war is one of the great disasters of all time for the United States. But Gene McCarthy hasn't been able to tap the unrest in the country. You have to be able to touch this uneasiness. . . ."

Kennedy then referred to the alienation abroad in the country. "It would be far more effective if McCarthy had been able to touch that spirit, if he went to what really troubles them, rather than getting on as he did, speaking about the war and being sort of a peace candidate."

Well, somebody asked him, why not support McCarthy instead of saying merely that he would support the nominee of the party? "President Johnson would like for me to come out for McCarthy," he said. "Then it would be a 'Kennedy-McCarthy movement.' I don't think it would do any good." Johnson, in a television interview with the three networks shortly before, had talked cryptically of "the Kennedy-McCarthy movement" in discussing McCarthy's campaign, always convinced that his archfoe, Kennedy, was behind it all. If Kennedy was able to rationalize for himself that his criticism of the Johnson Administration was not a vendetta, the President apparently never bothered engaging in that particular exercise.

Kennedy intimated that McCarthy, by needling him in public about his caution, was driving him and his supporters away. "McCarthy has hurt me by his taunting," he said, "and he hasn't helped himself. He's made it impossible for any Kennedy people to work for him. A couple of months ago, he probably could have put something together if he had done it right."

To some who heard him, there was in it an element of sour grapes, an implication that if the man in front had been Robert Kennedy, not Gene McCarthy, the challenge could have ignited something in the country. But that implication was blunted by Kennedy's own self-doubt; while pouring out all the rationalizations for staying out, he was obviously in torment over the inconsistencies of his position, especially when he talked of what might have been if the effort against Johnson had been "done right." Again he remarked: "I know that isn't very satisfactory, but I don't see that there is anything I can do about it." Well, then, he would support Johnson if nominated, was that right? He nodded hesitantly and added: "But not with enthusiasm." Then he smiled weakly. "It looks like a great year," he said.

In addition to the President's Vietnam policies, Kennedy was chagrined by what seemed to him to be Johnson's callous indifference to the plight of Negroes in the cities. "Johnson is on the side of repression now," he said when the discussion touched on what should be done about the big-city riots. Asked why, he blurted with some cynicism: "It's a political decision. There are more white people than black people." In spite of all he said to downgrade the feud, and to deplore the way it always was injected into the equation, it was there, and in moments of distress it spilled out nakedly.

As the backgrounder drew to an end, there was discussion about what could be placed on the record. Several of the participants, led by Paul Hope of the Washington *Star,* pressed hard for quotation of Kennedy's statement that he could not "conceive of any circumstances" under which he would run. Kennedy at first balked, asking, "Why do we have meetings like this?"—meaning why hold a backgrounder to enable a public figure to talk freely, then insist on placing what he has said on the record? But he had been around and he knew the game, so his remark was more a playful lament than a protest. It was decided that the reporters present could write that Kennedy had reiterated he had no plans to oppose Lyndon Johnson "under any conceivable circumstances." At this juncture, Mankiewicz broke in and suggested they make it "any foreseeable circumstances"—that phrase seeming less likely to foreclose the unexpected. Kennedy didn't disagree, but said he did not want whatever was printed to have the character of an announcement. He had considered making one, he said, but had abandoned the idea. Apparently he didn't care to leave the impression with Johnson, the party or the public that he had been seriously exploring the possibility of running. Rather, he wanted it to come off as casual.

One other incident occurred at the backgrounder that touched on an infinitely more critical factor in Robert Kennedy's political destiny than any of the semantics over which everyone had been quibbling. During the breakfast, Peter Lisagor of the Chicago *Daily News* passed a brief United Press International dispatch to Kennedy. It said that intensive Viet Cong attacks were under way in cities and towns throughout Vietnam. "Yeah, we're winning," Kennedy remarked, with sarcasm. Only when the Senator returned to his office did he get the first real inklings of what the UPI dispatch meant. The unforeseeable circumstances that would force him to reconsider his decision not to run were unfolding halfway around the globe; the Tet Offensive of January-February, 1968, had begun.

CHAPTER TWO

An End to Caution

THE DECISION TO RUN

THE MORNING AFTER Kennedy's breakfast backgrounder with the Washington political writers, Fred Dutton picked up the Washington *Post* and was astounded at what he read. He knew, from the meetings he had attended and from his own private talks with Kennedy, that the consensus clearly had been against the Senator's entry into the 1968 Presidential race, and that Kennedy was going along with it. But the way the phrase "under any foreseeable circumstances" seemed to come out in cold print—so flat and final—nevertheless flabbergasted the Washington lawyer-politician. He knew Kennedy really didn't feel all that definite.

Dutton went to the telephone and called the Senator. Kennedy said he too was surprised that in print the phrase seemed so final, that it indicated he had shut the door so completely on himself. What had occurred was one of those occupational differences in thinking and operating that separate reporters and politicians, a difference that few politicians ever really are able to grasp—to their endless dismay and, often, anger. Usually, when politicians talk, it is in the language of the hedge, the preservation of options, or so they hope. But when newsmen listen, they are trained to catch the hedge, to isolate it, weigh it, put it in perspective and carry the speaker to the full limit of its commitment. Thus, when Kennedy had said "under any conceivable circumstance," his news-minded—and nuance-minded—press secretary, Mankiewicz, had quickly moved in and suggested de-escalation to "any foreseeable circumstance." The second phrase left the door open a shade more than the first. But to the reporters, the second phrase still made the essential point—that unless something not anticipated happened, Kennedy was out of the race for 1968.

Kennedy was distressed about the way the story read, but more so at himself for having been so positive, than at the reporters. All of them, he subsequently learned, had placed essentially the same emphasis in their stories appearing around the country. He knew he had said what had been reported, and he even had agreed to have the phrase put on the record. So he was stuck with it.

The Senator might not have been so disturbed at himself had it not been for something else he read in the morning papers. The first rumblings about a step-up in Viet Cong activity throughout Vietnam—mentioned to him initially at the National Press Club backgrounder—now had grown to an avalanche of terrorist attacks on Saigon and a host of other towns supposedly "secure" for the South Vietnamese government. Conditioned by his own long disbelief of the Johnson Administration's rosy overcoatings and reinforced in this disbelief by what his brother Teddy, Nolan and other recent visitors to Vietnam had told him, Kennedy immediately cut through the official reassurances. He saw the burgeoning Tet Offensive for what it was—proof that American policy in that tiny half-country was bankrupt.

When the Tet Offensive came, American Marines and others were bracing for a major Viet Cong and North Vietnamese attack against the Central Highlands outpost of Khe Sanh, which challenged the main routes of troop infiltration from neighboring Laos. The general speculation was that the enemy was massing for a knockout blow that would make Khe Sanh a psychological American Dienbienphu—breaking the American will to fight just as the capture of that French outpost in 1954 had cracked resolve in Paris and led to the French pullout. The United States commander in Vietnam, General William C. Westmoreland, along with many others accepted the theory, and when the first Tet outbreaks occurred on January 30 and 31, they were taken as diversionary incidents subordinate to the Khe Sanh siege. Even when the American Embassy in Saigon was itself besieged, and held for several hours in a wild day of chaos and gunfire, the American penchant for verbalizing setbacks into proof of progress continued. The fact that the Embassy was retaken was hailed as a victory, ignoring the vulnerability that invited attack at all in the capital city of the Saigon government, supposedly the most secure in all of South Vietnam.

Claims were made of huge enemy losses, though the fighting somehow raged on; it was said the Viet Cong could not hold the cities they had successfully raided, as if that somehow proved there was internal security. Solace was taken from the absence of a public uprising in support of the Viet Cong—overlooking the reality that a huge infiltration and supply buildup had occurred without the supposedly "loyal" urban populace giving the government an inkling of what was to explode in its face.

Kennedy, already caught between his deep concern that Vietnam was a national tragedy and his pragmatic judgment that a political challenge would be suicidal and might only make matters worse, was plunged into even greater soul-searching. A columnist who had attended the National Press Club backgrounder interviewed him in his New York apartment three days later and found him already second-guessing himself. Kennedy asked his visitor whether the phrase "any foreseeable circumstances" was better than "any conceivable circumstances," and then he recited more political reasons why he ought and ought not to get in.

There were polls, Kennedy said, that showed him running well against the prospective Republican candidates in Nebraska, Iowa, Idaho and parts of Wisconsin. California, too, looked good for him, though it would not be as easy as some earlier polls had suggested. Kennedy said he thought he probably could win all or most of the primaries that still were open to him—at that point, all except New Hampshire. But then he switched again to the arguments against running: that his entry would tear the Democratic Party apart and would not bring him the nomination if party regulars held firm out of loyalty to the incumbent President. And he added one other interesting rationale: that his entry might bring down other Vietnam critics seeking reelection. Kennedy said a number of such Senate colleagues had advised him to stay out because his candidacy would put them on the spot, and in this context he specifically mentioned Senator McGovern.

Kennedy in the next days decided that he no longer could hold back on his Vietnam criticism. On February 8, with the Tet Offensive still going forward, he delivered his most devastating attack yet on Administration policy— without the customary backing away from Johnson as its architect. His book

To Seek a Newer World had just been published, and he had been invited to address the Book and Author luncheon of the Chicago *Sun-Times* in Chicago.

"Our enemy, savagely striking at will across all of South Vietnam," he said there, "has finally shattered the mask of official illusion with which we have concealed our true circumstances, even from ourselves." Earlier dreams of victory were gone, he said, and although the Viet Cong eventually would be driven from the cities and thousands killed, "they will nevertheless have demonstrated that no part or person of South Vietnam is secure from their attacks; neither district capitals nor American bases, neither the peasant in his rice paddy nor the commanding general of our own great forces."

The events of the Tet Offensive, he went on, "have taught us something. For the sake of those young Americans who are fighting today, if for no other reason, the time has come to take a new look at the war in Vietnam; not by cursing the past but by using it to illuminate the future. And the first and necessary step is to face the facts. It is to seek out the austere and painful reality of Vietnam, freed from wishful thinking, false hopes and sentimental dreams. . . . We will find no guide to the future in Vietnam unless we are bold enough to confront the grim anguish, the reality of that battlefield which was once a nation called South Vietnam, stripped of deceptive illusions. It is time for the truth."

The first illusion that had to go, Kennedy said, was the idea that U.S. response to the Tet Offensive constituted some kind of victory, as the Administration was claiming. The second was that the United States could win a war the South Vietnamese could not win for themselves. The third was that military victory at any cost was in the best interest of either the United States or the people of South Vietnam, and the fourth—in which Kennedy took dead aim on the Johnson Administration—was that "the American national interest is identical with, or should be subordinated to, the selfish interest of an incompetent military regime."

The Senator dismissed as "a pretension" the Administration claim "that the battle for South Vietnam is in reality a struggle for 250 million Asians—the beginning of a Great Society for all of Asia. . . . We can and should offer reasonable assistance to Asia; but we cannot build a Great Society there if we cannot build one in our own country. We cannot speak extravagantly of a struggle for 250 million Asians, when a struggle for 15 million in one Asian country so strains our forces that another Asian country, a fourth-rate power which we have already once defeated in battle, dares to seize an American ship and hold and humiliate her crew." (This reference was to the seizure of the American electronic spy ship *Pueblo,* with its crew of eighty-three, by North Korea on January 23. The Administration, despite pressures from Republicans in Congress to get tough, sought to dampen down the incident and seek recourse through regular diplomatic channels. Kennedy said later the mistake was to have not provided air cover for the ship in light of repeated protests of its presence by the North Koreans.)

Finally, Kennedy said, the United States had to disabuse itself of the illusion that "this war can be settled in our own way and in our own time on our own terms." There no longer was any prospect of a clear military victory, he said, so "we must actively seek a peaceful settlement" that would "give the Viet Cong a chance to participate in the political life of the country—not because we want them to, but because that is the only way in which this struggle can be settled." And this meant negotiating directly with the National Liberation Front as well as Hanoi.

Tet clearly had been the breaking point in Kennedy's reluctance to speak out for fear of adding to the Kennedy-Johnson feud talk. Yet there remained the hard political considerations of his actively challenging the President for renomination.

During his visit to Chicago, Kennedy expressed his concern about the war to Mayor Richard Daley. The mayor, the stereotype of the old politician, would never think about challenging an incumbent President, but he too was distressed about Vietnam. He mentioned to Kennedy the idea of a special Presidential commission to review the war, as one way possibly to shift American policy.

The alternative—opposing Johnson for the nomination—still seemed as fruitless as ever, to Kennedy as to his most trusted political counselors and friends. But Tet was working on them, too, or at least some of them. O'Donnell particularly was worked up about Vietnam and at one point he told another insider: "If he's not going to do anything about the war, I'm not going to stay with him." But O'Donnell said later that knowing the political realities as he did, he never pressured Kennedy to run. "I told him, 'If you want to run because of the issue, I'm with you. If you just want to get the White House limousines back, I'm against it.' " The younger, aggressive Senate staffers were champing all the more too. Adam Walinsky, whose devotion to Kennedy and belief in him had no peer, also told others he might quit if his boss did in fact support Johnson for reelection.

Ted Kennedy, recently back from Vietnam, agreed that the whole war was a mess and that Tet only confirmed that fact. But political pragmatism held the upper hand with him, and he still opposed his brother's candidacy. So did Ted Sorensen, who emerged as the toughest nut to crack of all the insiders. But many of the others who had been put off by whether deposing Johnson was "do-able" began to subordinate that consideration—they were no more convinced that it could be done—to the need for Kennedy to make the race as the strongest possible personal protest to Vietnam. Nolan, returning from a skiing vacation in Switzerland in February, felt Kennedy now could stay out only at his own political peril. He argued that the country would repudiate the war before November, Johnson would go down, and Kennedy would go down with him if he didn't cut himself loose. Some kind of settlement was due, Nolan insisted, and somebody was going to win the political allegiance of the great American middle road by taking the lead toward that settlement. Kennedy, of course, had done this in word back in February, 1966, but Nolan argued that he now had to do it in unmistakable deed.

Kennedy was in a frame of mind to listen, but he was a practical politician, and there remained persuasive reasons and persuasive voices that said "No." The characteristic round of individual conversations and exploratory phone calls started again. This time not only the Senator himself, but insiders who knew the extent and seriousness of his reassessment began—apparently at his urging—to call those they most trusted for their views. Joe Dolan, working out of Dutton's and Gwirtzman's downtown law firm to avoid publicity, began taking soundings by phone. Before long an underground flood of conversations was flowing in Washington and in other cities around the country, all focused on the same question: should he run? It was a miracle of modern-day secretiveness, in an era of wholesale leaking, that definite word of this extensive Kennedy agonizing and brain-picking did not blossom out in print until weeks later.

The reason may have been that life among the Kennedys went on as usual. Hickory Hill was aswirl with preparations for a massive television benefit for a Washington orphanage, Junior Village, and movie and television celebrities were brought in from all over the country to take part. The Saturday of the telethon and the Sunday after it, February 17 and 18, the house in McLean fairly crawled with show-business names and faces. In the midst of all the romping, politics was discussed on the run. Salinger produced a poll indicating McCarthy could get 40 percent of the Democratic vote in New Hampshire. "If it's true, you have to announce before the primary," Pierre told his host. Kennedy called Bill Dunphy, a knowledgeable former Democratic chairman in the state, and asked him about McCarthy's prospects. He was gaining, Dunphy said, but he doubted he'd do that well. Kennedy was wavering, but wasn't ready to jump. He pulled a letter from his pocket and showed it to Salinger. It was from Pete Hamill, a young newspaper columnist formerly of the New York *Post* and *Newsday,* who had gone to Ireland to write a novel. Hamill passionately denounced the war and urged Kennedy to run. That letter, Salinger said later, helped Kennedy over the brink.

But that weekend he still was holding on. On Sunday afternoon, the Senator and a few close advisers took refuge in the pool cabana to discuss his future. Ted Kennedy, Burke, vanden Heuvel and Goodwin were there, and the decision again was that he couldn't make the race. Goodwin was disappointed, and that night, driving to Dulles Airport with Kennedy, he told him: "As long as you're not running I think I'll go up to New Hampshire and work for McCarthy."

As the Tet Offensive stretched out into a full-scale and bloody siege of the old Vietnamese capital of Hue, however, Kennedy moved toward the edge of what many of the old hands still insisted was political suicide. To all who talked to Kennedy through late February, it was clear that, as Nolan put it, "in his gut" he wanted to run. Even those longtime associates who were not part of the soul-searching conversations could read the Kennedy mood of internal agonizing. "It was hard to reach him," his personal secretary, Angela Novello, recalled later. "I would give him messages and he wouldn't function." He was ready to go, but "from the neck up" it still didn't make sense. He needed something to push him over the edge, something more than Tet.

As February passed into March, he got it. The President's Special Commission on Civil Disorders, headed by Governor Otto Kerner of Illinois, issued a summary report on the root causes of the previous summer's riots. The Commission fixed the primary blame on "white racism" at the core of big-city ghetto life—an indictment that met with stony and conspicuous silence at the White House. Repeated efforts by White House correspondents to elicit some comment, or to find out if the Administration intended to implement commission recommendations for costly crash programs of ghetto rehabilitation, got nowhere. In most cases of Presidential commission findings, the White House is used as a launching pad to assure maximum exposure, often with the President himself releasing the report and praising it. But this time it was nearly a week before any official or semiofficial Administration position appeared, and then it came in a statement from Vice President Hubert H. Humphrey expressing clear reservations about the prime conclusions and sweeping recommendations.

For Kennedy, this was the last straw—or so he convinced himself. He had been able to live with the rationalization that he could differ basically with Johnson on Vietnam and yet support him for reelection on domestic issues. "This means," he told those privy to his soul-searching, "that he's not going to do anything about the war and he's not going to do anything about the cities, either." The Kerner Report was leaked on Thursday, February 29, and then released on Friday. The very next day, Saturday, March 2, there were informal discussions at Hickory Hill, during which Kennedy took a phone call from Jesse Unruh in California. The results of his poll, taken by John Kraft, gave Kennedy 46 percent of the total, against only 21 percent for a pro-Johnson slate, 13 percent for McCarthy, and the rest undecided. To the catalysts of Tet and the Kerner Report, there now was a persuasive political input. Around this time, there also was a call from Al Lowenstein in New Hampshire, telling Kennedy that McCarthy would do very well in New Hampshire and would beat Johnson in Wisconsin in April. Kennedy said he didn't believe it, but he listened.

Developments in the Republican Presidential race also had an impact on him. On February 28, Michigan's Governor George Romney had jolted his party and the nation by abruptly withdrawing with two weeks still to go in the New Hampshire primary, in which he was opposing former Vice President Richard M. Nixon. Romney, after more than a year of campaigning marked by self-damaging obfuscation, faced such overwhelming defeat that he pulled out to deny Nixon the victory and to enable the party's liberals and moderates to rally around somebody else—obviously New York's Governor Nelson A. Rockefeller. Kennedy immediately recognized that Romney's action only made Nixon look all the stronger, and it convinced Kennedy more than ever that Nixon would be the Republican nominee. To all the Kennedys, the thought of Richard Nixon in the White House, eight years after having been beaten by John Kennedy, was bitter to contemplate.

On Sunday, March 3, the informal talks continued. Mankiewicz, arriving at National Airport from out of town, called Kennedy at Hickory Hill to remind him to watch a television documentary on the Senator's pet urban rehabilitation project in New York, Bedford-Stuyvesant. Kennedy informed his press secretary that talks were going on about his candidacy—Mankiewicz' first hard knowledge that it had reached the decision stage. "Want to take my vote?" he asked his boss. "Yes," Kennedy said. "Go in everything after Wisconsin," Mankiewicz told him.

That same day Kennedy phoned Fred Dutton, who had just gone to California on law business, and asked him to rush back east to talk about the situation. Dutton caught an early plane and was in Kennedy's Senate office the next morning, Monday, March 4. The Senator had a specific mission for Dutton. Robert Kennedy himself was ready to go, but his brother Teddy still had doubts. Would Fred, as an old 1960 hand who earlier had shared Teddy's reservations about the wisdom of making the race in 1968, talk to him? It seemed strange that one of these very close brothers would send an emissary to intercede with the other, but Robert Kennedy apparently wanted to free Teddy from the personal bind in this critical decision. Dutton agreed, and arranged to see Ted Kennedy in his Senate office the next morning, Tuesday, March 5.

Although Dutton walked into Ted Kennedy's office that morning prepared to marshall all the arguments against the war and for making the race, they proved to be unnecessary. "Bob's just about made up his mind to run," Ted told him at the outset. "The thing now is to make sense of it." He proceeded to discuss the pros and cons, from a political viewpoint, neither minimizing the difficulties nor exaggerating the strengths. Then he and Dutton went to his brother's office, where over a three-hour lunch they had the first hard talk, not about the why and why not of it, but the when and how. There was specific discussion of announcing the candidacy two days later—on Thursday, March 7—five days before New Hampshire's Democrats were to go to the polls. Thus it was abundantly clear that Kennedy had decided to run well before the results of that primary vote revealed Johnson's vulnerability. He had "decided," but as was his nature, he still was exploring and, in conversations with others in subsequent days, he continued to say he was "thinking about" running. But the plans went forward from this point on.

Kennedy, after having weighed the situation, decided against the March 7 announcement. Further reports from Bill Dunphy in New Hampshire indicated McCarthy was coming on, and Kennedy didn't want to appear to be pulling the rug out from under him—to be acting "ruthlessly." Besides, he hoped McCarthy would do well, for both tactical and personal reasons. First of all, it just might work as a restraint on Johnson in his Vietnam policy. Second, it would demonstrate Johnson's weakness to Democratic pros. Third—and perhaps most important to Robert Kennedy in his obsession that his opposition would be regarded as nothing more than a personal vendetta—it would show that the Democratic Party was split, regardless of what he did. And there also was the possibility that McCarthy's quiet campaign style was best suited to demonstrate Johnson's vulnerability. "Had Robert Kennedy gone into New Hampshire with sixty newsmen trailing after him," Steve Smith said later, "it would have been a quite different kind of campaign."

The best move, they decided, was to stay out until after New Hampshire. A ragtag "RFK in '68" Committee there was asked to discourage any write-in for Kennedy. A half-page advertisement in the Manchester *Union Leader* specifically made the plea on the day before the primary, Monday, March 11. It was cleverly worded to imply that Kennedy was requesting that those who would have written in his name vote instead for McCarthy. He had not, of course, made such a request, choosing instead to remain quiet and neutral until the March 12 vote.

There were no illusions, even then, about Kennedy's chances for the nomination. He probably couldn't win, but he had to get in because of the war, and the other lesser catalysts. With the March 7 announcement of candidacy rejected, talk turned to the need to take a last reading after the New Hampshire results were in. Robert Kennedy was scheduled to appear at a United Jewish Appeal dinner in New York the next Tuesday, March 12—the date of the primary; the group at first talked of meeting in Steve Smith's apartment there later that night. But on second thought, the results would still be coming in and the picture might not be clear. It would be better to wait a day, so the date was set for Wednesday, March 13. The secret meeting broke up, with all participants aware that the cause would not be served if word of it leaked out.

Not only was most of the press and the public in the dark about the seriousness of Kennedy's intentions at this point; so also were some key members of his Senate staff, including those young aides who wanted most for him to run. But the flow of bodies in and out, and the phone conversations, soon gave them the sense, in that last week, of what was going on. Kennedy, in his agonizing, had not felt it necessary to call on most of them. He knew how they felt, and what he needed at this juncture was to test his own feelings against hardheaded opposition, not eager persuasions that he was right in wanting to run.

This did not mean that the Kennedy office was doing business as usual. The staff was constantly discussing Presidential politics now—Kennedy, McCarthy, Johnson, New Hampshire. Several days earlier, Joe Dolan and Frank Mankiewicz, looking down the road, had begun to talk between themselves about the California primary and the need to file a petition of intent there. The deadline for listing a delegate slate was April 5, seemingly still a long way off. But Dolan noted a quirk in the California law that required, thirty days before the deadline, that a committee of registered Democrats file papers indicating intent to run a slate. It wasn't necessary to identify the candidate. If their boss were to run, Dolan and Mankiewicz decided, he would have to be in the California primary. So, without bothering to advise him, they went ahead and—"just for protection"—found some California Democrats willing to file the intent papers.

On Tuesday, March 4, Dolan decided he'd better mention the quirk in the California law to Kennedy, but he said nothing of the effort he and Mankiewicz already had started. Kennedy repeated that he hadn't made up his mind but, just in case, maybe the intent requirement ought to be met—on the quiet. He suggested that Jesse Unruh be asked to handle it. So Joe called off the first group and phoned Unruh, who said he would get seven California lawyers to form the committee.

The next day, March 5, nothing was heard in Kennedy's office until fifteen minutes before the filing deadline. Then Mankiewicz' phone rang. It was a UPI reporter in Sacramento. "What do you know about the five little old ladies who took out the Kennedy papers?" he wanted to know. Frank expressed bafflement—with good reason. At this point, there had been no sign of Unruh's seven lawyers. The ladies, the UPI man said, had insisted they were acting on their own; that if they had asked anybody in authority, they would have been told not to do it, so they just went ahead. Five minutes later, Unruh's seven lawyers rushed breathlessly in, just under the deadline. If the little old ladies hadn't acted, and if the Unruh group had been ten minutes later, Kennedy would have been barred from the California primary. "The vaunted Kennedy machine," Mankiewicz observed wryly later, "almost blew it."

The question remained about how to handle McCarthy. He had come to Kennedy back in November to say he was running if Kennedy was not, and he had to be told now about Kennedy's change of plans. If Kennedy was going to stay out for another week allegedly in deference to him, he wanted McCarthy to know about it. "I asked my brother to tell McCarthy," Kennedy told David Wise, a magazine writer, later. "My brother decided, on his own, that he didn't want to tell him in the last days of the primary campaign. So we both talked to Dick Goodwin, who was working for Gene McCarthy, and asked him to tell McCarthy, and he did, on the day before the primary." Ted Kennedy reported later that he was to tell McCarthy "that Bob was very much considering running," that he phoned Goodwin and asked him to do it, and that Goodwin called back on Sunday, March 10, or Monday, March 11, to say he had passed the word in McCarthy's room at the Sheraton-Wayfarer in Manchester.

According to Goodwin, Robert Kennedy asked him to tell McCarthy only that he was "thinking of running." Goodwin declined to be a formal emissary but agreed to pass the message. McCarthy hadn't expected Kennedy to enter the race, Goodwin said later, but he took the news calmly. There never was an attempt at a real accommodation between the two, then or later, according to Goodwin. "You can't have an accommodation between two guys who want to be President," he said. "And after New Hampshire, McCarthy sure wanted to be President. At first, Kennedy didn't think McCarthy had a chance, but after New Hampshire he always felt McCarthy could be formidable. He always understood the people around McCarthy would get an emotional involvement in him and wouldn't be moved. But he felt he could beat him and knock him out."

By this time, even without word from Kennedy, McCarthy could have read the signs for himself, though the secrecy wall showed surprisingly few cracks. On Thursday, March 7, Kennedy had an off-the-record private lunch in his office with CBS newscaster Walter Cronkite, just back from Vietnam. Cronkite had broken with established network tradition of newscaster neutrality by criticizing the war directly on his television show. At the lunch Cronkite heard the Senator's strong views on Vietnam and told him: "If you feel that way you ought to run." Kennedy replied: "Give me five reasons." But from what he already had said, it was clear to Cronkite that Kennedy was on the brink. Then Kennedy astounded the newscaster by asking him whether he was a Democrat. Cronkite said he was an independent. What Kennedy had in mind was for him to run for the Senate in New York. Cronkite said it was out of the question, that he always had maintained a nonpartisan posture as a newsman, and always would.

Later that afternoon, when Cronkite got back to his studio to prepare for his evening show, he found that the network had a report from CBS newsman John Hart, saying Kennedy had been reexamining his position and was about ready to announce his candidacy. Cronkite was appalled; it might seem to Kennedy that he had breached the Senator's confidence. He phoned Kennedy, told him through Mankiewicz about Hart's report, and his own quandary. Cronkite did not want to break Kennedy's confidence, but neither could he, in good conscience, censor his own network or reject the accurate work of an honest and enterprising reporter. Kennedy, engaged on the Senate floor, sent back a message through his press secretary: "The Senator says that you can say he's considering running for President just as Walter Cronkite is considering running for the Senate in New York." The Hart segment was included in the Cronkite report that night, Thursday, March 7. But surprisingly it didn't cause much stir at the time. And Cronkite's budding political career, which apparently had been conceived in Kennedy's fertile political mind but nowhere else, died on the vine—to Cronkite's great relief.

In the Senate that same day, Kennedy threw out a signal for all those who were alert enough to see it. He led a blistering attack on Johnson's Vietnam policies, the like of which—in intensity and in number of colleagues participating—had not been heard in that sedate chamber for a long time. Senator J. W. Fulbright triggered it with an expression of concern over reports that the Pentagon had asked the President for 206,000 more American troops in Vietnam. Before any such action was taken, he insisted, the Senate should be consulted. Kennedy rose to second the demand.

"Every time we have had difficulty over a period of the past seven years," he said, "over the period during which I was in the executive branch, and since I left the executive branch, the answer has always been to escalate the conflict. It has always been to send more troops. And at the time we sent the larger number of troops, or increased the bombing, we have always stated that there would be light at the end of the tunnel, that victory is just ahead of us. The fact is that victory is not just ahead of us. It was not in 1961 or 1962, when I was one of those who predicted there was a light at the end of the tunnel. There was not in 1963 or 1964 or 1965 or 1966 or 1967, and there is not now. . . .

"Moreover, there is a question of our moral responsibility. Are we like the God of the Old Testament that we can decide, in Washington, D.C., what cities, what towns, what hamlets in Vietnam are going to be destroyed? Is it because we think it may possibly protect the people of Thailand, the people of Malaysia, the people of Hawaii, or keep certain people out of Texas or California or Massachusetts or New York? Or do we have that authority to kill tens and tens of thousands of people because we say we have a commitment to the South Vietnamese people? But have they been consulted, in Hue, in Ben Tre, or in the other towns that have been destroyed? Do we have the authority to put hundreds of thousands of people—in fact, millions of people—into refugee camps for their protection, or should these decisions be left to them?"

Kennedy compared the American presence in Vietnam with the movements of the Germans and Russians into Poland and the Baltic States in World War II "because they needed them for their own protection . . . they needed them as a buffer. I question whether we have that right in this country." He talked of corruption in the South Vietnamese draft and hit Johnson directly with it: "When this was brought to the attention of the President, he replied that there is stealing in Beaumont, Texas. If there is stealing in Beaumont, Texas, it is not bringing about the death of American boys."

Among the alert reporters who immediately saw the implications of this Senate outburst was Haynes Johnson of the Washington *Star*. The next morning, Johnson called Angie Novello and asked if he could see the Senator sometime in the next few days. Within several minutes, she phoned back and told Johnson to come right up to the Capitol Hill office. For about ninety minutes, Kennedy unburdened himself about the quandary in which he found himself. He couldn't live with this war, and yet if he ran against Lyndon Johnson everybody would say it was only the vendetta and his own ruthlessness. But Kennedy had no doubt by this time that the President was on the ropes. He said McCarthy would do well in New Hampshire and would beat Johnson in Wisconsin the next month. After the session, the reporter scribbled his dominant impression in a small memo book. The entry said: "March 8, 1968—Spent 1½ hrs. with RFK. Certain he will run."

This was on Friday, and on Saturday, March 9, Kennedy left Washington for the first time in nine days to attend a fund-raising dinner in Des Moines for Governor Harold E. Hughes of Iowa. He was accompanied by his research aide, Edelman. It was a routine affair, except that afterward there was a gathering of Kennedy, Hughes and three other Midwestern Democratic governors— Robert Docking of Kansas, William L. Guy of North Dakota and Warren E. Hearnes of Missouri—plus some Democratic Congressmen and other party hangers-on. There were about thirty individuals in the room in all, and in a sense it was a private meeting. But with that many, and no security, it was one of the most public private sessions on record. The governors were in a grumpy mood, and didn't care who knew it. They spilled out all their gripes against President Johnson, confirmed each other's views that he was in grave political trouble in the Midwest, and expressed fears that not only would he go down in 1968, but many other Democratic candidates with him. All four were facing reelection or were considering running for some other office, and it was personal as well as professional concern. Some criticized the war and others said the party ought to find another candidate. Nobody, in the recollection of participants, flatly suggested or urged Kennedy to make the race, nor did he say he intended to do it. But the mood certainly was such that it could only persuade him the more that he ought to run.

From Des Moines the next morning, Sunday, March 10, Kennedy and Edelman, joined by John Seigenthaler, flew out to Los Angeles to visit Cesar Chavez, the Mexican-American leader of the California migratory grape pickers who was about to end a twenty-five-day hunger strike. Ed Guthman joined them in Los Angeles, and the party of four flew to Delano, the Chavez movement's headquarters in the Central Valley north of Bakersfield. Kennedy long had been struck by the plight of the California migratory workers and impressed by their determination and by the grit and guts of their young leader. For his part, Chavez respected Kennedy as he did few other politicians; and two weeks earlier, when doctors became concerned about the health of Chavez, they called Kennedy in the hope he might persuade Chavez to break his fast. But that step hadn't been necessary, and now Kennedy wanted to be there when the fast finally was to end voluntarily.

Chavez, lean and weak, first met the Senator in a small chapel with a picture of John F. Kennedy on the wall, and then the two walked in a procession to an outdoor Mass in a public park. Kennedy sat next to Chavez and at Communion the fast was broken, with Chavez and the others eating bread. From all accounts, the long and time-consuming odyssey to Delano was not political, but rather something Kennedy felt he must do. On return to Los Angeles, Kennedy phoned Unruh and told him he probably would run. But that was all. The news certainly came as no surprise to Unruh by now. Kennedy's old Justice Department aide, John Nolan, even then was in California to help lay plans, at Kennedy's instruction, for contesting the California primary.

Sunday, March 10, was notable for one other development that was to cement Kennedy's now all-but-certain decision to run. Word leaked out that the White House had in fact been asked by the Pentagon generals to send the additional 206,000 American troops to Vietnam. This news, in O'Donnell's view, was what finally persuaded Kennedy to run. Kennedy called Sorensen, one of the last remaining holdouts, in New York.

"I really think I have got to go," he told his late brother's alter ego. Sorensen was unmoved. "I don't see how you are going to get a majority of the delegates," he told Kennedy.

What happened next is a subject of lingering controversy between the Kennedy and Johnson partisans, because the course of Kennedy's deliberations now reached over into the enemy camp, into direct contact with the White

House and the President. The sources of information about what was said to whom and when, and what the reactions were, divide into two channels, one pro-Kennedy, the other anti-Kennedy. But on the essentials there is sufficient agreement to record this much about the episode:

In the same telephone conversation, Sorensen advised Kennedy that he was going to the White House the next afternoon, Monday, March 11, for a meeting with the President, at Johnson's request, conveyed through an aide and mutual friend. Sorensen, at the request of John Criswell of the Democratic National Committee and of his White House friend, had interceded earlier with a group of New Hampshire Democrats determined to run a primary slate for Kennedy, and had persuaded them to withdraw. The President, Sorensen was told, wanted to express his appreciation. In any event, Sorensen would drop by Kennedy's Senate office Monday morning, before going to the White House.

When Sorensen arrived, Kennedy, who had flown in from Los Angeles, began again to marshal the reasons he felt he had to run. Again, he placed heaviest weight on the deteriorated situation in Vietnam, while acknowledging to his pragmatist visitor that he was aware he probably would lose. "I know when I look at it realistically that what you say about the arithmetic [the number of delegates he could hope to corral at the convention] is right," Kennedy told him.

According to Sorensen, he told Kennedy of his intention to propose a Vietnam review commission to Johnson, and Kennedy insisted later that the idea was Sorensen's own. Arriving at the White House, the former Presidential counsel was ushered into Johnson's office. Sorensen passed the word to the President—certainly with Kennedy's approval—that the junior Senator from New York was seriously considering a challenge for the Democratic Presidential nomination. According to the Kennedy camp version, Johnson then asked Sorensen if he had any suggestions—suggestions presumably to avert Kennedy's entry, although no one later would admit that such was the thrust of the question. Noting the turmoil in the country over the Vietnam War, Sorensen proposed the appointment of a blue-ribbon commission to take a fresh look at the war and to report to the President. He observed that Johnson had taken that approach in dealing with the epidemic of big-city riots, appointing the Commission on Civil Disorders under Governor Kerner.

According to Kennedy later, Sorensen "did not get this idea from me, was not representing me, did not propose me as a member and did not tie the idea in any way as to my prospective candidacy." Johnson told Sorensen, Kennedy said later, that "a similar idea had been advanced by a political leader"—obviously Mayor Daley of Chicago, who had broached the same idea to Kennedy in February—and that the President "welcomed the idea and wanted Mr. Sorensen to suggest names of the commission members."

After the Sorensen-Johnson meeting, which lasted about two hours and also touched on politics in general, Sorensen left the White House and phoned Kennedy. The Senator, still grappling with himself over which way to go, discussed with Sorensen and with his brother Teddy what he ought to do about the commission idea. Still maintaining even in his innermost circle of confidants that his only motivation was ending the war, not personal ambition or the issue with Johnson, Kennedy reasoned aloud that such a commission if carried through could justify his staying out of the race. "He really had a fetish about this idea that he would be breaking up the party," O'Donnell recalled later. "To suggest a commission on the war was the height of childishness. Commissions don't run wars. But he had to shut that last door."

The rest of Monday and much of Tuesday, March 12—primary day in New Hampshire—were given over to discussions of the commission idea in the top echelon of the Kennedy strategy board. Only a few of the most trusted associates were told about it—Steve Smith, O'Donnell, Fred Dutton and one or two others. There were phone calls between Kennedy and Dick Daley, and the White House and Daley, the latter still hoping to stem Kennedy's candidacy with the commission scheme. And all the while, the pressures continued to build on Kennedy from others who wanted him to run and who didn't know, or weren't sure, how close to the point of no return he already had advanced.

Among those pushing hardest now was Walinsky, and he reached out for allies wherever he could find them. Although Kennedy was the Senator from New York, his ties to, and his interests in, Massachusetts continued to be strong. One of the newspapers that continued to come to the Kennedy office was the Boston *Globe*, whose young and aggressive Washington bureau chief, James Doyle, had just returned from Vietnam. Walinsky and Mankiewicz had read Doyle's poignant and often devastating dispatches on the chaos and futility of the war—particularly the recent drawn-out siege of Hue—and when Mankiewicz met Doyle at the Gridiron dinner the previous Saturday night, he had urged the reporter to come to the office Monday to talk to Kennedy. Doyle did so. While he was waiting to see the Senator, Kennedy came out of the office with Walinsky, walking to the Senate for a vote. He spied Doyle and asked him to go along. As they walked, Walinsky encouraged Doyle to tell Kennedy what he had seen and learned in Vietnam. The reporter, thoroughly dismayed at the war, spoke emotionally of the destruction and the hardships imposed on the people, on the children, of South Vietnam by the American presence and power. Kennedy, who had been in high good humor until then, just listened, until Doyle finally turned to him and said, "What are you going to do?" Kennedy looked at him, as if Doyle had just hit him below the belt. There was no answer for several moments, and then, he replied, simply, "I don't know."

They were at the foot of the steps to the Senate chamber now, and Kennedy turned to Doyle and said, "Wait here, will you?" While he was gone, Walinsky asked Doyle flatly: "Do you want him to run?" Doyle, hopeful that Kennedy could somehow find a solution to Vietnam, said he did. "Then when he comes back, just talk about Vietnam," Walinsky told him. "Don't ask him any questions, just talk." Walinsky obviously felt that Kennedy, on the brink, needed encouragement and justification, not needling.

When Kennedy returned, Doyle resumed his account of his brief Vietnam tour, Kennedy again mostly listening as the three of them walked back to Kennedy's office in the bright, clear March day. They took an elevator to the third floor and as they got off, a group of other reporters was standing there, waiting. As they came up, Kennedy turned to Doyle and asked, "What do you think we should do over there?"

"Get out and salvage some dignity," Doyle answered. "It's a military problem. We're losing the war militarily."

"No," Kennedy replied. "It's a political problem."

"Well, at any rate," Doyle said, "all of us should do all we can to end it."

Kennedy smiled. "Yeah, I got your message earlier, Jim," he said.

The message was coming through just as loud and clear from others, but there remained the matter of the Vietnam commission. Kennedy continued to weigh it, considering possible members, that day and the next—Tuesday,

March 12—as New Hampshire Democrats went to the polls to choose between McCarthy and Johnson.

The President, who had not campaigned at all, had unwisely left his fortunes to a trio of the most incredible political bunglers yet assembled—Governor John W. King, Senator Thomas J. McIntyre and Bernard J. Boutin, a former subcabinet bureaucrat and odd-jobs man in the Kennedy and Johnson Administrations. At the outset they set the stage by predicting that McCarthy would be lucky to get 10 percent of the vote. Then they circulated numbered cards committing Democratic voters to vote for Johnson—cards that were to be signed and sent on to the White House. And when McCarthy seized on the drive as evidence of Johnsonian arm-twisting, they countered by charging his Vietnam criticism gave aid and comfort to the enemy. King went so far as to say there would be "dancing in the streets of Hanoi" if McCarthy won in New Hampshire. A few days before the primary, Boutin was suggesting that anything less than 40 percent of the vote would be a disaster for McCarthy. The Titanic Trio had botched the Johnson campaign, but they were bright enough at the end to be aware that McCarthy might do well, and to try to blur the results. As it turned out, McCarthy averted a disaster, even by Boutin's standards.

At 5:30 P.M., before the polls closed in New Hampshire, Kennedy boarded a plane at National Airport and went to New York to attend the United Jewish Appeal dinner. On Kennedy's arrival newsmen asked him if the New Hampshire vote would change his position as a noncandidate. He told them he wasn't changing his mind "at this time." But another development already had occurred that was pushing Kennedy to the edge: testimony by Secretary of State Dean Rusk before the Senate Foreign Relations Committee, glossing over the Tet Offensive and signaling a stand-pat policy in Vietnam by the Administration. By the time Kennedy reached his New York apartment in the late evening, it already was clear that McCarthy was giving Johnson a scare in New Hampshire's popular voting, and possibly was beating him soundly in the race for twenty-four national convention delegates. After late supper at "21" with Schlesinger and vanden Heuvel, Kennedy tried to call McCarthy to congratulate him, then sent a telegram.

By the next morning, when Kennedy boarded his return flight for Washington, the results had firmed up. It was a resounding upset, the dimensions of which and the public reaction to which Kennedy had not anticipated. McCarthy had won 42.2 percent of the Democratic vote, to 49.4 percent for the incumbent President, and had picked up twenty of the twenty-four delegates. (The winning Johnson delegates were the Titanic Trio—King, McIntyre and Boutin—and the party's state chairman, William Craig.) There also were some McCarthy write-ins in the Republican primary, won overwhelmingly by Richard Nixon, raising the possibility that McCarthy actually had outpolled Johnson in the total New Hampshire vote. Kennedy, who had asked his diehard supporters in the state not to write in his name, found that they had listened to him. He got 600 votes, or a paltry 1.1 percent. On the Republican side, Nixon had an easy time with Romney out and with only a feeble, amateurish write-in for Rockefeller going against him. Rockefeller himself had not yet moved, and although he now was expected to enter the race, he had tried to dissociate himself from the late and foredoomed write-in.

In the previous week's timetable of the Kennedy inner circle, this night—Wednesday, March 13—was supposed to be the kickoff. The word had been passed to the other insiders to be at Steve Smith's apartment at 1030 Fifth Avenue in the late afternoon, ostensibly for the presenting of any eleventh-hour briefs to Kennedy against entering the race, and then to end the suspense. Some thought they were being summoned to decide whether Kennedy should run or not; those closest to him knew that the decision was all but made, and that their business that night would be to decide not what, but how. Still, Kennedy *was* hanging by his fingertips, not wanting to drop off the precipice until the last possible moment; so the whole thing still could be aborted. Those few who knew of the Sorensen-Johnson talk and the commission proposal were more aware of this fact than the others. But they also knew best that it would take a totally unforeseen development at this juncture to deter Kennedy.

When the flight from New York landed at National Airport around noon, Sam Donaldson of ABC News accosted Kennedy and asked him how McCarthy's showing in New Hampshire affected his own position. For some reason—even his closest confidants contended later that they still could not fathom it—Kennedy blurted it out: "I am actively reconsidering the possibilities that are available to me, and I imagine that other people around the country are doing the same." And to another newsman, he admitted he was "reassessing the possibility of whether I will run against President Johnson" for the Democratic Presidential nomination.

After weeks of agonizing in secrecy; after hours of discussion about how entering the race against Johnson would compound his "ruthless" image; after the decision of a week earlier not to announce before the New Hampshire primary so as to avoid appearing to be "ruthlessly" pulling the rug from under McCarthy; after all this, Kennedy had blown it. Only hours after McCarthy had held a victory press conference in Manchester and as he was flying south to taste the sweet fruits of acclaim from his supporters in Washington, the man who could have been the anti-Johnson candidate in New Hampshire, but who had backed away from it, had jumped in and stolen the spotlight. It was—few critics had trouble finding the word for it—ruthless and insensitive to McCarthy's hour of triumph. (McCarthy, on his arrival at National later in the day, was asked if he would welcome Kennedy's entry into the race. "Well, I don't know," he said coldly. "It's a little bit crowded now.")

Not only did Kennedy say he was reassessing his position, but he enumerated for reporters five reasons for getting into the race—if he did get in: the party already was split on the war, so his candidacy could not be blamed; Rusk's testimony indicated no change in Administration policy on Vietnam; New Hampshire's results suggested Richard Nixon would be the Republican candidate, offering no hope of a Vietnam change; Administration silence on the Kerner Report indicated no movement on the cities crisis; his own feelings were compelling him to act. "Maybe he didn't know he was doing it," a member of the inner circle said later, "but he was announcing right there."

The statements electrified Washington. Calls and telegrams began to pour into Kennedy's Senate office urging him to run, or not run. The press swarmed around outside his door. Some reporters who knew him well were called in. One of them was Bruce Biossat, the gently perceptive columnist for Scripps-Howard's Newspaper Enterprise Association. Kennedy didn't say he had fully decided to run, but all he did say clearly pointed that way. McCarthy's showing in New Hampshire, he told Biossat, had demonstrated that "the divisions in the Democratic Party are already there and I can't be blamed for creating them." For the first time, Kennedy now was able to argue—or to rationalize—that the Kennedy-Johnson feud had nothing to do with it. It was a point he was to make frequently in private and in public in the next week, as all his fears about accusations of "personal vendetta" and "ruthlessness" materialized at the worst possible time and in the worst possible way.

When Biossat asked Kennedy what he could say to the inescapable charge that he had waited for McCarthy to demonstrate Johnson's vulnerability, and then tried to hijack his anti-Johnson, antiwar movement, Kennedy tossed it aside. "This is nobody's movement," he said. "I have to do it my way and McCarthy has to do it his way." The dissenting youth, the disaffection resulting from the war, the opposition to Johnson didn't belong to McCarthy or to anyone else, he went on. It was clear that Kennedy still could not see McCarthy as a viable threat to topple Johnson, did not think McCarthy had the drive, the compassion or the public following to get the job done—or to explain to the country adequately why it had to be done.

Although Kennedy discussed the war, he dwelt passionately on the Kerner Report, and on Johnson's bland reaction to its explosive findings and recommendations. Not only was Johnson going to stand pat in Vietnam, but in dealing with the cities too. He disagreed basically with Johnson in foreign policy, and now it was the same on the domestic front.

Finally, Kennedy plugged in the political arguments. He mentioned the Des Moines meeting with Hughes, Docking, Guy and Hearnes on the previous Saturday night; Mayor Daley, he said, had predicted Johnson would lose to Nixon in the fall. (But Daley later the very same day publicly reiterated his support for Johnson, and Governor Hughes in Iowa said he would "hate to see the Democratic Party torn apart" by a challenge from Kennedy.)

If the airport interview with Donaldson did not burn the last bridges, Kennedy was about to do so. Walter Cronkite, learning of the development, called and invited Kennedy to appear on his mass-exposure evening television news show. Kennedy earlier had agreed to do so if he decided to run, and in the late afternoon he drove to the CBS studio in northwest Washington, where he taped an interview for use at 6:30 P.M. Cronkite questioned him from New York.

Meanwhile, in Steve Smith's New York apartment, the insiders had begun to arrive as early as 4 P.M. There, then or later, were Ted Kennedy and Smith, Salinger, Sorensen, O'Donnell, Dutton, Dolan, Schlesinger, vanden Heuvel, Burke Marshall, Nolan, Jerry Bruno (the Kennedy advance man), Tom Johnston, Dave Burke, Milton Gwirtzman, Barrett Prettyman, John Burns, Jack English. Others came in and out in the course of the evening, including sisters Jean Smith and Pat Lawford. Notably, the Young Turks—Walinsky, Greenfield, Edelman and others who long had wanted him to run—were not there.

Talk about running started informally, with Sorensen the only significant participant still making a case against it. By not running, he argued, Kennedy would retain the latent support of the Democratic Party in the nominating process. If the convention became snarled, he said, it would turn to Kennedy as a loyal member who in spite of severe reservations had not bolted. Most of those present acknowledged, as another participant put it later, "that the chances were marginal or slim that he could pull it off, but an informal consensus was reached that he should go." They believed, by this time, that raising the antiwar issue was enough, if the country thus could be rallied to force a change in the Vietnam policy. Of Sorensen, another insider said later with some bitterness: "He didn't disagree that the cause was right. He just didn't agree it was worth losing for." Steve Smith's own reaction, summed up later, seemed to express what most felt. "I always had an ambivalence about it," he said. "What commended it was the spirit: the hell with the consequences, you have to do it. I could feel both sides, but the decision was essentially his, and all we could do was help him with it."

There was some talk of Kennedy supporting and working for McCarthy, and several spoke favorably of the idea. But it never really got off the ground. To men who considered themselves guardians of Robert Kennedy's stature, that step seemed small, half-way, even self-demeaning. It was taking a shot at Johnson without risking or accomplishing anything. Most believed, with Kennedy, that McCarthy was a paper tiger. If it was to be done at all, their man would have to—and should—do it himself.

It was in this atmosphere, with the insiders ready to recommend that he go ahead if he felt he must, that they awaited Kennedy's arrival from Washington. Sometime before 6:30, a call came from Washington telling them to be sure to watch the Cronkite show; they gathered around the Smiths' television set as Cronkite introduced his tense, solemn guest and asked him about his "reassessments."

As he had done for his National Airport interviewer earlier, Kennedy ticked off the reasons he felt compelled to consider the race. Again, the argument that New Hampshire had removed the deterrent of the Kennedy-Johnson feud led the rest. "I was reluctant to become involved in this struggle because I thought it might turn into a personal conflict between President Johnson and myself," he said, "and that the issues that I believe strongly in and which I think are being ignored at the moment would be passed over."

New Hampshire "has demonstrated that there is a deep division within the Democratic Party," he went on. "One of the major reasons that I didn't want to become involved earlier was because I thought that I might be the instrument of dividing either the country in a way that it would be difficult to put back together [or] dividing the Democratic Party in a very damaging way."

There were the other reasons too: Secretary of State Rusk's stand-pat testimony on the war; the President's noncommittal reaction to the Kerner Report; the likelihood that Nixon would be the Republican nominee, thus threatening a continuation of the status quo no matter who won.

Kennedy said he hadn't discussed his reassessment with McCarthy but wanted to do so "before I finally make up my mind as to what I should do in the future. . . . I want to have a chance to talk with him about the future and about what he's committed to, the policies. He's committed to bringing about this change I think that both of us are interested in, and I would like to talk to him about what he feels that perhaps I can contribute."

To Cronkite and others listening, the last remark suggested he might just support McCarthy rather than run himself. Cronkite asked him about it, but Kennedy would say only that "I want to decide that and what I will do after I have a chance to assess it all." But those closest to his thinking already knew he was not seriously considering throwing in with McCarthy, about whom he had marked reservations. Still, there was the definite tactical problem of how to work with McCarthy, and hopefully, to nudge him out of the picture early. Actually, shortly after the TV taping, Kennedy met with McCarthy in Ted Kennedy's office and told him he probably would run. The meeting was curt and cool, especially on the part of McCarthy, who told him he could do as he pleased, but he—McCarthy—would continue to pursue his own course.

In Steve Smith's apartment, the insiders watched and listened to the Cronkite interview with rapt attention. What they were viewing, they knew at once, was their man's announcement for the Presidency. It wasn't final and it wasn't official, but it was clear to them that now only some momentous development could deter him; any more talk by them at this meeting about whether or not he should run would be just so much wheel-spinning. "We all realized right then," one participant said later, "that the whole thing—what we thought we were there to decide—had really been decided by Bob on national television." Said another: "The ball game was over, right there."

In this frame of mind, the insiders sat down to dinner. Talk turned from what to do to how to do it, but in a chatty, convivial and informal way. The working session melted into a Kennedy-type dinner party of bright and witty talk. Latecomers arriving, such as O'Donnell, whose plane from Boston had been diverted to Philadelphia because of fog and rain, were told any more debate was "academic." When O'Donnell walked in, Smith told him: "He's already gone and done it."

According to the pro-Kennedy account later, there was a phone call during dinner, which Steve Smith took. It was from Sorensen, who had left the apartment early to keep an appointment in Rochester. When he arrived at LaGuardia airport for his flight, a message awaited him. The White House had phoned for his list of participants on the proposed Vietnam commission. Sorensen called the Smiths' to tell Kennedy, who was en route from Washington, but the Senator had not yet arrived. So he left a cryptic message that Kennedy would understand but other, lesser, insiders would not grasp.

Robert and Ethel Kennedy reached the Smiths' at about 8:30 or 9 P.M., when dinner was nearly over. He was greeted as "the candidate" amid much applause, congratulations and warmhearted ribbing. He himself was in a much-relieved mood and Ethel was buoyant as usual. They sat at the dinner table for a while, and in half an hour or so the gathering broke up into small discussion groups. The Kennedy brothers, Smith and one or two others went into the den, where they began to talk strategy, the matter of a formal announcement and the one item of unfinished business that might yet abort the whole decision—the Vietnam commission.

Robert Kennedy and his top associates began to phone Democratic political leaders around the country, still taking soundings and Kennedy still saying only that he was "thinking" about running, or that he would "probably" run. One of those called was Dick Daley in Chicago. Again according to the pro-Kennedy account, Daley urged Kennedy not to declare. "I still wish the President would change his Vietnam policy," Daley reportedly agonized over the phone. "You wouldn't feel this way." And in one last effort to avert the split, he said he would call Johnson directly. Shortly afterward Daley called back. According to the pro-Kennedy account, the Mayor said the President had told him: "I'm all for this commission. I'm waiting for Sorensen to give me the names." Daley then is said to have suggested that Kennedy contact the new Secretary of Defense, Clark Clifford, presumably at Johnson's behest.

Robert Kennedy, still clinging to the precipice, immediately began to scramble back. Maybe the cooler heads would prevail after all. He asked brother Teddy to phone Clifford, Washington's most firmly established honest broker, who also happened to be the man best positioned to act as a link with Johnson. Clifford was a Johnson insider long before he joined the Administration as McNamara's replacement, but he also was close to the Kennedys, having served on occasion as a personal lawyer and legal counselor.

It was then about 11 or 11:30 P.M., so Teddy called Clifford at home. According to the pro-Administration account, Teddy told the Secretary only that his brother wanted to talk to him about Vietnam, but Ted Kennedy said later the Vietnam commission definitely was mentioned then. At any rate, Clifford readily agreed to the meeting and said he would come to Robert Kennedy's office. No, Teddy said, Bob would rather make it at Clifford's, in the Pentagon. An appointment was set for 11 o'clock the next morning.

Only a few of the Kennedy insiders in Smith's apartment knew of this last development—the Kennedys, Smith, O'Donnell, Dutton, one or two others. Sorensen was phoned in Rochester and he made arrangements to be in Washington the next morning.

Meanwhile, the discussion groups made their first rough explorations of the scope and problems of the prospective campaign. Three main groups were working, at the informal direction of Ted Kennedy and Smith, with Robert Kennedy drifting from one to another, mostly just listening. There were no appointments and no orders. One group concentrated on New York, with home-staters Burns and English involved in that; another discussed the situation in California, the key primary state, with Dutton, a one-time aide to former Gov-

ernor Pat Brown there, importantly involved. Ted Kennedy led the examina-
tion of the situation in the other states. The New York and California groups
were in the living room, the rest around the dining room table. "Let's see,"
Teddy said, "there's California over there, there's New York over there. That
leaves forty-eight for me." His aide, Burke, opened a black loose-leaf notebook
containing Kennedy supporters by states. The first page, marked "Alabama,"
was blank. "Will the candidate reconsider?" somebody asked. "The candidate"
still had not said in so many words that he was one, but those who knew nothing of the Clifford meeting had no doubts—and no reason to have any.

Eventually the talk among the insiders turned to the formality of announcing. Logic seemed to dictate Monday, March 18; most thought Kennedy should
stay out of the Nebraska primary, whose deadline was to be Friday, March 15.
It was a Republican state, and the one in which John Kennedy had done most
poorly in 1960. By holding off the announcement until Monday, Kennedy could
finesse that one and still be on the record in plenty of time to stay in the Oregon
primary, whose deadline for filing affidavits of disavowal was March 22. But
Sorensen thought Kennedy, if he was going to run, ought to go into Nebraska.
It was Sorensen's home state, his brother Phil, a former lieutenant governor,
could help handle things, and he believed (rightly, it later proved) that Kennedy
would do well there. Besides, after Kennedy's statements on the Cronkite show,
it would be hard to persuade the Nebraska Secretary of State, who drew up the
list of likely candidates, to keep Kennedy off. The insiders, communications-
minded, knew that Saturday morning might be the best time; the announcement would produce the main news for the nation's big-circulation Sunday
newspapers and would draw an invitation from one of the major news panel
shows on television Sunday. But nothing firm had been decided on the date by
the time the last of the conferees—Salinger, Dutton, O'Donnell, Nolan—left
the apartment at about 1 A.M.

Sorensen met Kennedy in his Capitol Hill office the next morning and drove
with him to the Pentagon for the 11 o'clock meeting with Secretary of Defense
Clifford in his spacious "E" Ring office overlooking the Potomac River. What
happened in that room on the top floor of the world's largest office building,
within view of the Washington Monument, remains an issue of dispute between the Kennedy and Administration forces. That the participants were Ken-

nedy, Clifford and Sorensen, and that the matter of the Vietnam commission and its members was the main order of business, is agreed. But concerning the emphasis placed on various points, and what Kennedy wanted, it is necessary to put down each side's version.

The first point of dispute is whether Clifford, before Kennedy and Sorensen arrived, knew what they were going to talk about. Kennedy, in discussing the matter three days later, after the news of the meeting had been leaked by the White House, said specifically that Clifford "had previously been told of the [commission] idea by the President," and Ted Kennedy said later that he had told Clifford of the topic. But the Administration version holds that Clifford had not been told—although he phoned Johnson Friday morning from his home presumably to talk to him about the impending meeting. That he would enter into such a meeting without being advised by the President of the previous discussion with Sorensen—indeed, that Johnson would explore the matter to this point without confiding in his Defense Secretary and old close friend and adviser—defies credulity. But, in any event, that is the Administration version.

The Administration story also insists that not until the moment Sorensen walked into his office with Kennedy did Clifford know that it was not to be a two-man session between him and the Senator. Much is made of this point in the pro-Johnson version, because it is said to underline the point that Clifford had not yet been plugged in on the commission idea. If he had known Sorensen was accompanying Kennedy, it is argued, he certainly would have told Johnson, who in turn quite likely would have advised Clifford that he already had seen Sorensen, and why.

At any rate, both sides agree that Kennedy opened the discussion with a recital of his deep concern about the way things were going in Vietnam; that the Administration was on the wrong course there; that the conduct of the war was the overriding issue in the country; that his own deep feeling was moving him to challenge Johnson's renomination. The Administration version holds that Kennedy did not say that if Johnson acted in accordance with his wishes he would not run, but that was the definite impression left with Clifford. Kennedy three days later said for the record that "although I made it clear to Mr. Clifford I was contemplating a race for the Presidency largely because of the failure of our Vietnam policy, the three of us explored the commission idea, in Mr. Clifford's words, 'not as a proposal by either of us,' not as a help to me but as a means of helping President Johnson move away from our presently disastrous course in Vietnam." But, Kennedy went on, he did make clear that "if it were more than a public-relations gimmick, if both the President's announcement of the commission and its membership signaled a clear-cut willingness to seek a wider path to peace in Vietnam, then my declaration of candidacy would no longer be necessary. Ending the bloodshed in Vietnam is far more important to me than starting a Presidential campaign."

Sorensen then outlined the plan for a commission to reassess and re-evaluate Vietnam policy. According to the Administration scenario, he listed as prospective members Edwin Reischauer, the former ambassador to Japan; Roswell Gilpatric, former Deputy Secretary of Defense; Carl Kaysen, former National Security Council assistant to President Kennedy; General Lauris Norstad, former NATO commander; General Matthew Ridgway, former United Nations Commander in Korea; Kingman Brewster, president of Yale University; and Senator Kennedy. In the Kennedy version, the Senator said he would "willingly serve on such a commission, but I did not insist on that, and I stated that I should not be chairman."

According to the Administration account, Clifford noted there was only one Senator on the list, whereupon Kennedy and Sorensen suggested Senate Majority Leader Mike Mansfield—like Kennedy, a Democrat. "You don't have any Republicans," Clifford noted next. His visitors agreed and suggested Senators George D. Aiken of Vermont and John Sherman Cooper of Kentucky. In all of this, it is pointed out, Clifford did not push for the additions, just noted the imbalance. In the later Kennedy version, there was no suggestion that Mansfield, Aiken and Cooper were additions to the original list.

Sorensen then is said to have emphasized that as part of the arrangement the findings of the commission would be reported to the President and made public, and when the commission was announced, the President would say he was disturbed, that he felt the country might be off course on the war, that he wanted others to take a look, and he would be bound by their findings. To this, the Administration version holds, Kennedy himself objected, saying it would be impossible for any President to make such a statement, but that he should say he would give the greatest possible weight to what the commission found. Sorensen then was said to have suggested that Kennedy be chairman, which Kennedy—as he himself emphasized later—also rejected as not necessary. The Administration script says Kennedy wanted to be a member; Kennedy said later he would "willingly serve but I did not insist on that."

The meeting broke up on the understanding, Kennedy later said, that what had been discussed "should be kept completely confidential," and that Clifford would check with the President and advise Kennedy. Kennedy and Sorensen were insistent that they have an answer that day, according to the Administration version.

After the two visitors left, Clifford phoned the President and made an appointment to see him that afternoon, shortly before 4 P.M. When Clifford entered Johnson's office and laid the Kennedy proposal before him, the President's reaction was, in the Administration version (the only available one, in this case), "instantaneous and very clear." He found the whole idea to be totally unacceptable, for several reasons. First, it would look like a deal no matter how it was dressed up; second, it was difficult for the President to think of anything that would be more comforting to Hanoi; third, even to give the appearance of turning over to any such commission a decision of this nature would be a clear usurpation of Presidential authority; fourth, it would create havoc among other Vietnam critics like Senate Foreign Relations Committee Chairman Fulbright and Senator McCarthy; fifth, it was clear from the proposed membership of the committee, even before it met, what its findings would be, since all the men proposed had been openly critical of the Vietnam policy.

Clifford, at about 5 P.M., called Sorensen, who was in Kennedy's office with the Senator. When Sorensen took the call, he asked Clifford to wait until Kennedy could pick up an extension. The Secretary then relayed the President's firm and unequivocal rejection to both men. The Administration version is that there was no further discussion; Kennedy said later that Clifford had given as two reasons "the undesirability of my being on the commission, and the fact that President Kennedy had not named a similar commission after the Bay of Pigs." Kennedy said he had asked Clifford whether the fact he (Kennedy) didn't insist on serving on the commission would change the President's position. Clifford said it would not. "I also pointed out that President Kennedy had indeed appointed a commission after the Bay of Pigs," Robert Kennedy, who served on that commission, added.

According to Kennedy, Sorensen had later talks with "the White House" that confirmed the rejection, "although different reasons were given." It was then, Kennedy said, that it "became unmistakably clear to me that so long as Lyndon B. Johnson was President, our Vietnam policy would consist of only more war, more troops, more killing, and more senseless destruction of the country we were supposedly there to save."

In all the speculation later about the precise moment when Kennedy finally decided, once and for all, to run, the heaviest evidence is that this was that moment. "Bob wanted to stop the war but he also wanted to preserve his long-term political future," one of the insiders said later. "If he could have achieved both by staying out, he would not have run. But when Clifford told him, 'No, we can't work it out,' the die was cast. Bob had said he was going to run; when Johnson slammed the door he had no alternative. He was trying to keep his options open long after he really had lost them. Everything was in motion by then." O'Donnell, against the commission idea from the start, said later: "It didn't matter. He was committed. There was no road back."

That night, Kennedy went home to Hickory Hill for a routine dinner with New York weekly newspaper editors, in town for a convention. McCarthy already had reacted to Kennedy's first "reassessment" statement by announcing that day he would enter two more primaries, in Indiana and South Dakota. The message from him was clear too: he was not going to be scared out of the race. Thus it was a final resignation in Kennedy's mind, rather than a sudden flash of purpose, that he spoke of three days later when he said of Johnson's rejection on Thursday, March 14, of the Vietnam commission: "That night I decided to run for President."

CHAPTER THREE

The First Days

ALL-OUT AGAINST JOHNSON

AT ABOUT 2 o'clock on Friday morning, March 15, the phone rang in the Alexandria, Virginia, home of Haynes Johnson of the Washington *Star*. It was Angie Novello, Kennedy's private secretary, calling to say the Senator was going up to New York that morning and would Haynes like to go along. The reporter, reading the invitation correctly, said he would. He met Kennedy, virtually unnoticed at National Airport, and they caught the Eastern Airlines shuttle to LaGuardia—the second shuttle, an old and noisy Super-Constellation. In the rush of getting aboard there was no time to talk, but then they settled in near the rear of the plane. Kennedy took the window seat of a three-place row, with Johnson next to him and Jack Newfield of the *Village Voice*, who had been at Hickory Hill Thursday night with the New York editors, on the aisle. Kennedy was nervous. He sat staring out the window, tapping his front teeth with his fingernails—a nervous characteristic of his late brother.

"Well," Johnson finally asked, "how does it stand?"

Kennedy turned to him. "I'm going to announce tomorrow morning," he said flatly.

It hadn't been an easy decision, he told his companion. "A lot of people who were my friends will be distressed," he said. One of them was former Secretary of Defense Robert S. McNamara, whom he already had informed. And there were others. "I think I'm in by myself," he said, although he knew most of his close friends were sticking with him, even if they thought he was wrong to try it.

Kennedy tried to keep his monologue on the issues, on his concern about the war and the way the country was going at home, especially in the cities. But inevitably it became personal, though he seemed to be defensively guarding against that turn. He spoke of the things that had been started in the Kennedy Administration and that in his view had stalled under his brother's successor; of the crises in the Kennedy Administration, and of his own role in dampening them down—Berlin, and the Cuban missile affair; of the partial nuclear test ban treaty; the beginnings of Vietnam. And finally he said: "My brother suggested Lyndon Johnson to be President of the United States. All that is involved inside me. I represent something more than myself. President Kennedy began something—that's what would be ruthless and callous, if I ignored that."

He had considered going in before the New Hampshire primary, even entering it belatedly, he said, but "it was a dilemma and a problem either way. Any way I did it was not going to be accepted." Going into a state where his brother had begun his own bid for the Presidency might have seemed an appeal to the past. And there was the business of the feud with Johnson.

"My own feelings were to run," he said, still in a grim mood, "but for all of these other reasons I didn't. It's a much more natural thing for me to run. When you start acting unnaturally, you're in trouble. At least now I can react normally to events and issues." And in the same vein: "I have to deal with myself, and that's a problem. I have to balance that off, I have to decide whether I'm going to act like a hypocrite or stand up for what I believe."

When the plane landed at LaGuardia, they were met by Tom Johnston, Kennedy's New York Senatorial aide. The group drove along the East River to Steve Smith's apartment on the clear, brisk day, and Kennedy's gloom lifted en route. By the time they arrived at the Smiths' Kennedy was ready to get moving. He took off his coat, rolled up one shirtsleeve and began to work the phone, calling friends and political associates. Surprisingly, some of the closest insiders—Pierre Salinger was one—got the final indication only then. He called his brother Teddy, John Nolan, James Wechsler of the New York *Post* and a string of Democratic politicians. Of one, a woman, he asked: "Do you love me in March as you did in December?"

The rest of the day, scheduled long before, required his attendance at a round of humdrum political and civic meetings on Long Island. At one of them, a gathering of more than a hundred housewives and neighbors at the Kings Point home of Mrs. Anita Richmond, a woman asked him: "What can I do to help you?" Kennedy told her, as reporters scampered for the nearest available telephones: "Well, you can help me in this great effort I'm going to undertake."

That did it. The afternoon papers bannered the news. Some of them also carried other remarks that were a prelude to his later elaborate rationale for entering after New Hampshire. At Mrs. Richmond's, when the subject of his relations with President Johnson came up, he complained that Johnson always had interpreted his disagreement on issues as a personality struggle. "Maybe it is personal animosity," he said at one point, "because he feels that I think John Kennedy should still be President of the United States. That is what President Johnson has been saying." And as to why he had not jumped into the race earlier, he said, "If I had entered the New Hampshire primary, they would have thought I was trying to divide the Democratic Party. I feel strongly about the party and I don't want to divide it."

He was in now, all right, but tradition and the pulling power of nationwide television required that he save his formal declaration for the next morning, in a proper setting, properly transmitted by the networks into millions of American homes. So he went on to Garden City for a businessmen's "fall guy" luncheon meeting and listened to stale gags while such delights as "Grilled Texas Bird, A la Mo" and "Boston Baked Baines" were served.

At one stop, Kennedy found himself sitting at the head table next to Leonard W. Hall, the former Republican National Chairman and campaign manager for Dwight D. Eisenhower. A New York radio station had an open microphone on the table and broadcast this bit of conversation:

Kennedy: "Do you think I'm crazy?"

Hall (muffled by noise): ". . . What will McCarthy do?"

Kennedy: "He's going to stay in. So it's going to make it tough . . . (noise) . . . It will make it much tougher."

Hall: "Are you going the primary route?"

Kennedy: "Yeah."

On the flight back to Washington, Kennedy discussed the day and the days ahead with Haynes Johnson, then went home, where a typical Hickory Hill work-play dinner party was in progress, with record player blaring. The task of writing his important statement of candidacy still was to be resolved. Adam Walinsky, as was his wont, already had worked up a draft, but Kennedy wanted some of the older speechwriting hands—particularly Sorensen and Schlesinger—to get in their licks. Also present were brother Teddy, Steve Smith and Fred Dutton—all on the political team, and Walinsky came later. Ethel also had invited some of their social friends—the George Stevenses, Susie Markham, the Dave Hacketts and some others of the Washington set. The dinner table was set for about a dozen, but several sat on the fringes and still others, who came expecting a hard business session, already had eaten.

After dinner the important work at hand was conducted in a small study just to the left of the entrance hall, while in the spacious living room beyond, the music continued. It was, as quite often at Hickory Hill, a bizarre mix. Among the guests was Al Lowenstein, the New Yorker who had launched the dump-Johnson effort in the summer, tried to get Kennedy, General Gavin, Kenneth Galbraith and George McGovern to lead it, then turned to McCarthy. He flitted between the workers in the study and the players in the living room, and there was a momentary verbal clash when somebody asked him, "Who are you with?" Said one of the participants in the drafting group later, "Bob enjoyed it. He liked to put a bunch of strange fish in a bowl and see how they reacted."

In the study, there were gathered that night three generations of successful speechwriters—all prideful of their work and their generation. There was Schlesinger of the Stevenson era, Sorensen of the John Kennedy era, and Walinsky of the Robert Kennedy era. In the ensuing discussion of what the candidate should say the next morning, the clash of temperaments and of styles was a delight for bystanders to behold. Walinsky, seeing his draft being ground to pieces in the discussion, lobbied for it diligently among the participating insiders. He captured Kennedy's speaking style best of the three; Sorensen's suggestions seemed to some to be just a bit pompous. At one point, Kennedy observed about a Sorensen high-flown phrase, "I can't say that." Finally, sometime after midnight, the candidate turned to Schlesinger and asked him to "reconcile these drafts." Walinsky groaned audibly, and hung in. The three worked on, with Schlesinger finally going home and leaving the speech to the other two. The announcement statement, in style, was mostly Sorensen's—deliberate, a bit lofty, repetitive for emphasis. It was, according to one insider, the last major speech he wrote for Robert Kennedy. Schlesinger submitted some drafts after that, but most of the written campaign oratory fell back to those who had been more closely associated with the candidate in the Senate—Walinsky, Jeff Greenfield and others of the young entourage referred to by the older hands as "the Kiddie Corps."

There was one other order of business that Friday night—telling McCarthy, then campaigning in remote Green Bay, Wisconsin. It was decided that the use of Goodwin as an emissary was not enough this time, so Ted Kennedy caught a plane to Chicago and, after much difficulty, on in the small hours into northern Wisconsin. He arrived at McCarthy's hotel at about a quarter to three in the morning. The situation at any hour hardly was conducive to cordiality and profitable political talk, and it turned out to be simply a protocol visit—"a kind of goodwill mission," McCarthy called it later in the day. "I don't know whether it was meant to come to anything in particular, but it was a courtesy call and I appreciate very much the effort that Senator Kennedy made," McCarthy said. "But there was no offer of any deal from him to me and certainly no response on my part except what I've said here that I don't intend to make any deals."

McCarthy's comment on the Ted Kennedy mission left the impression that the younger Kennedy had arrived unexpected in the middle of the night and had roused McCarthy from slumber. According to Ted Kennedy, however, McCarthy not only knew he was coming, but a McCarthy staff plane was sent to Chicago to pick him up. "I had indicated that if it was worthwhile to visit on the question of how Bob intended to run, the kinds of things he would stress, I would be glad to go up," Ted said later. "I said I wasn't interested in going just for the ride, and he [McCarthy] indicated it would be worthwhile." When the younger Kennedy got to Green Bay, he said later, he spent forty-five minutes to an hour discussing his brother's plans in what essentially was a one-sided conversation. Then he flew back to Washington.

Robert Kennedy's campaign for the Presidency started officially at 10 o'clock the next morning, Saturday, March 16, in the ornate old Senate Caucus Room where his brother had announced his own Presidential candidacy on January 2, 1960, and where Robert Kennedy earlier had gained national prominence on his own as a dragon-slaying inquisitor for the Senate Labor Rackets Committee. It was a cloudy, warm day for March, and the room was packed with reporters, photographers and an overflow crowd of Kennedy backers and tourists. A few minutes before he entered, old John Kennedy hands came in and scrounged for standing room. One of them was Schlesinger, with his usual mixed look of milquetoastishness and importance. As he entered, he was roughly bustled to the rear of the room by an unknowing cop. Ethel Kennedy, bright-eyed but a bit tense in a blue coat, white gloves and white net stockings, led nine of her brood—all except the baby, Douglas—to seats at the front. At the rear of the family entourage was fifteen-year-old Joe, Kennedy's oldest son, tall in a dark blue pinstripe suit but gangling and adolescent. All the boys needed haircuts in the worst way, and otherwise also resembled their father.

Kennedy, accompanied by Mankiewicz, came in at the appointed hour, in a blue suit and red-and-blue figured tie fastened with a gold PT-109 tieclasp. He was alert and serious, but smiled as friends caught his eye. Behind him as he walked to the microphones were Newfield and staff aide Edelman. He seemed impatient to get started, and even before the usual wave of demanding photographers had subsided, he began reading—firmly, with no trace of nervousness—from a large black notebook with the words typed in oversize letters. "I am announcing today my candidacy for the Presidency of the United States," he said. "I do not run for the Presidency merely to oppose any man"—Kennedy's obsession with the personal vendetta charge was obvious in this opening line—"but to propose new policies. I run because I am convinced that this country is on a perilous course and because I have such strong feelings about what must be done that I am obliged to do all I can. . . . I run because it is now unmistakably clear that we can change these disastrous, divisive policies only by changing the men who make them. For the reality of recent events in Vietnam has been glossed over with illusions. The report of the Riot Commission has been largely ignored. The crisis in gold, the crisis in our cities, the crises on our farms and in our ghettos, all have been met with too little and too late. . . ."

Kennedy had words of praise for McCarthy's "remarkable New Hampshire campaign," but they were not uttered without a purpose of his own. That effort "has proven how deep are the present divisions within our party and country. Until that was publicly clear, my presence in the race would have been seen as a clash of personalities rather than issues. But now that the fight is one over policies which I have long been challenging, I must enter that race. The fight is just beginning, and I believe that I can win."

Fully aware of the new round of "ruthlessness" talk, Kennedy sought to blunt it. He told of his younger brother's flight to Wisconsin to see McCarthy and said he had made clear through Ted Kennedy "that my candidacy would not be in opposition to his, but in harmony." (At this, muffled guffaws buzzed through the press section.) His aim, Kennedy went on, was to "both support and expand his [McCarthy's] valiant campaign" to bring a change in Vietnam policy. "Taking one month at a time, it is important now that he achieve the largest possible majorities next month in the Wisconsin, Pennsylvania and Massachusetts primaries. I strongly support his effort in those states and urge all my friends to give him their votes."

Kennedy said he would enter the California primary on June 4 "in the belief, which I will strive to implement, that Senator McCarthy's forces and mine will be able to work together in one form or another." (More snickers from the press section. It was apparent, even then, that this was a diplomatic way of saying he hoped McCarthy would see the handwriting on the wall by the time of the primary in California and bow out, leaving the challenge there to Kennedy.) The new candidate also would enter the primaries in Nebraska (May 14) and Oregon (May 28) because, he said, state laws required that his name be placed on the ballot. "But in no state," he sought to stress, "will my effort be directed against Senator McCarthy."

Finally, he came back to the matter that bothered him so much: "My decision reflects no personal animosity or disrespect toward President Johnson. He served President Kennedy with the utmost loyalty and was extremely kind to me and members of my family in the difficult months which followed the events of November, 1963. I have often commended his efforts in health, education and many other areas; and I have deep sympathy for the burdens he carries today. But the issue is not personal; it is our profound differences over where we are heading."

The declaration was short and designed to get him off and running—but not, if it could be managed, over the prostrate forms of Johnson and McCarthy. That hope, the ensuing questions from the press soon showed, was an idle one.

Almost at once, he was hit with it. "There have been speculations," somebody asked, "that this is opportunism on your part, that McCarthy had the courage to go into New Hampshire, and now after his success—" The question trailed off. Kennedy tried lamely to combat it. "As I said, I've spoken on these issues and these questions for a number of years, and how I feel about them," he said. "I felt, and I think it was generally accepted, that if I had gone into the primary in New Hampshire or if I had done well in the primary in New Hampshire, it would have been felt at that time that it was a personal struggle. It would have been written in the press as a personal struggle. Every time I have spoken on Vietnam over the period of the past several years, every time I have spoken on what I think needs to be done as far as the cities are concerned, it's been put in the context of a personal struggle between myself and President Johnson. Therefore we would get away from what the issues are, which divide this country. I think the New Hampshire primary established that the division that exists in this country, the division that exists in the Democratic Party, are there, that I haven't brought that about, that what had brought that about is what President Johnson—the policies that are being followed by President Johnson. . . . I'm not asking for a free ride. I'm putting my candidacy and what I believe that I can do for the future of this country and what needs to be done in this country to the American people."

Well, somebody else asked, wasn't his candidacy likely to so divide the party's opposition to Johnson as to "virtually assure the nomination of President Johnson" and "make it easier for a Republican to win in November?" Kennedy argued that by running he actually would "broaden" rather than narrow the protest against Johnson's policies. "I think Senator McCarthy has certain strength," he said. "I think I add. I think I have certain strength. And together, that we add to the opposition that presently exists."

It was a logic that did not go down well. A reporter inquired why he had rejected "the alternative of putting your strength behind Senator McCarthy." Kennedy said he didn't think "just supporting an individual" could deliver the strength his own candidacy could muster. It went on like that, with Kennedy hard pressed to make his case, to the end. It was doubtful, when it was over, that he had made believers of many in the room who still thought the timing and manner of his entry was "ruthless," no matter how he rationalized it.

These first minutes of Robert Kennedy's Presidential campaign had put him on the defensive, and he obviously was glad to have them behind him. The crush began at once—the crush that had been Kennedy's constant companion since the assassination of his brother, and that had grown in intensity and passion in the ensuing four years, three months and twenty-three days since that act of madness in Dallas. He, Ethel and the children pushed through the hordes of well-wishers inside the Caucus Room, out into the corridor, down the winding marble staircase and outside to the waiting cars. At every turn, the hands were outstretched to shake his, or to touch him. He was, in a real sense, home among these Washingtonians; they did not display quite the squealing fervor that inundated him elsewhere around the country. But even here, it was electric. Now, and until the final journey eighty-five days later back to this same spot on Capitol Hill and on to the quiet hillside across the Potomac where his brother, John F. Kennedy, waited in eternity, he was a public possession. In the course of these next eighty-five days, Robert Kennedy was to write a chapter in the history of American political campaigning unprecedented in its pace and intensity, and in the passions it aroused. For a time, the campaign was so frenzied that the personality of the candidate was overshadowed by his image of a compulsive figure carried along by the tide and by the sheer force of the Kennedy political legend of inevitable success. But as the campaign took shape, and was shaped by events, he emerged increasingly as a distinct political personality, not divorced from the family epic, but with an identity of his own beyond the fact that he was the brother of a martyred President embarked on a mission of family restoration. In these first days, it was his intensity that came through; later, as the campaign downshifted a bit, a warmth and relaxed good humor surfaced that surprised those who had seen only the early intensity or had heard mostly of the "ruthlessness."

Trailed by reporters, Robert Kennedy raced to National Airport for the next regular flight to New York, where he was to march in the St. Patrick's Day parade. He was a declared candidate, and this was officially his first campaign trip, but so frantic had been the days leading up to his final decision to run that the vaunted Kennedy machine and logistics were practically nonexistent. He boarded with a planeload of regular Saturday passengers, bewildered by all the hoopla and the sight of a Presidential candidate in their midst.

While the candidate was en route to New York, the state of campaign unpreparedness was being addressed in a meeting in Ted Kennedy's Senate office. Among those present were Steve Smith, Sorensen, Dolan, Dutton, Mankiewicz, Walinsky, Edelman, vanden Heuvel and Gwirtzman. Brother-in-law Steve took charge as a kind of de facto campaign manager, doling out obvious jobs. Steve and Sorensen would come to Washington to oversee money, policy and strategy; Teddy and Ken O'Donnell would look into problems of the nonprimary states; Tom Johnston would move down from New York, too; Edelman and Gwirtzman would research issues; Burke would be Teddy's aide, which he was anyway; Bruno, as always, would be a key advance man; Walinsky and Greenfield would handle the bulk of speechwriting. None of this was startling. Some primary states were assigned, willy-nilly. The traveling campaign team had to be put together. Mankiewicz would handle press on the road, assisted by Hugh McDonald, a young former reporter working in the Senate office on a Congressional fellowship. As a Californian, Mankiewicz also would be plugged in later on policy and operations in that state's primary, with Teddy's highly capable press man, Dick Drayne, replacing him on the road. The meeting stretched through the afternoon. That night at Hickory Hill, the candidate personally asked Dutton to be his traveling aide-de-camp, a task he was to perform in the next months with energy, patience, dependability and good humor. He first had come to the attention of John Kennedy as a chief aide to Governor Pat Brown in California. Dutton's talents as a political diplomat, conciliator and troubleshooter led to his appointment as secretary of the Cabinet and then as Assistant Secretary of State for Congressional Affairs, in which post he demonstrated his abilities as a liaison man. He was a natural as the one campaign lieutenant who could convey Robert Kennedy's instructions to a varied staff of old and young hands with a minimum of friction and ruffled egos. He and other old 1960 aides worked at their own insistence as volunteers, on expenses only. They did not want to be lost amid the staff—the younger, salaried team members. They thus staked out from the start a special relationship with the candidate. He was the boss, but they worked more with him than for him, an arrangement that was mutually beneficial, if not always organizationally tidy.

During the afternoon session in Ted Kennedy's office, Mankiewicz took a phone call from a *Newsweek* magazine correspondent. What about a proposal to the President that Kennedy would stay out of the race in exchange for appointment of a public commission on Vietnam? Mankiewicz said later this was the first he even knew about the idea. Others also soon had it—*Life, Time,* and by the next day CBS. It is the Kennedy camp's contention that the White House leaked the story Saturday to paint Kennedy as a blackmailer, meddling in high government policy. The White House never has commented on how the story got out. The fact the news magazines didn't publish until Monday gave the Kennedy forces a little time to handle the matter, but not much. Kennedy was advised as he carried on his first official Presidential campaigning in New York City Saturday afternoon, and the decision was to sit tight.

In Wisconsin, McCarthy maintained his blasé attitude toward Kennedy's entry, but not without a few penetrating barbs. He recalled that he had taken on Johnson when "a lot of other politicians were afraid to come down onto the playing field. They were willing," he went on, "to stay on the mountain and light signal fires and bonfires and dance in the light of the moon, but none of them came down . . . I walked alone. They weren't even coming in from outside; just throwing a message over the fence. . . ."

And in Washington, at a businessmen's conference, President Johnson took the barest public note of the Kennedy candidacy. "These are days when we have to take chances," he told the group wryly. "Some speculate in gold—a primary metal—and others just speculate in primaries."

The Kennedy campaign started, appropriately, in an Irish bar on West 48th Street, called Charley O's, where a festive St. Patrick's Day party was in full swing when Kennedy arrived, as usual, late. It was to have been his annual blast for the New York press. Bagpipes squealed and drums rolled as he pushed in, photographers forming a wedge around him, and up to the bar. There, the guests put down their Irish coffees and crowded around, offering congratulations. "Thank you, thank you very much," he said in his customary monotone. He hopped onto a table, a green and white button that said BOBBY O pinned to his jacket, to extend St. Patrick's Day greetings to all, and to be kissed and given a sprig of shamrocks by a pretty Aer Lingus stewardess. The Donegal Pipe Band of New York, ten stout men in green and black kilts, lined up and marched him out to the parade on Fifth Avenue. They played "The Wearin' of the Green," "Minstrel Boy to the War, Has Gone," "Back to Donegal" and John Kennedy's favorite, "The Boys of Wexford."

The parade was not all it should have been for a Presidential campaign kickoff. He was Irish and he was at least an adopted New Yorker, but he also was against the war, and as he moved along Fifth from 48th to 86th, there were a few boos and catcalls in nearly every block. The cheers outnumbered them, however, and he kept going. He shook hands with the newly appointed Archbishop of New York, the Most Reverend Terence J. Cooke, on the steps of St. Patrick's Cathedral and waved to Mayor John Lindsay at the 63rd Street reviewing stand. There were cries like "You'll never make it, ya bum!" from the crowd, and near the end a husky young man carrying a St. Francis College banner turned on him angrily, protesting his presence in the parade. "I swear to God, if he didn't have twenty cops around him, I'd punch him right in the mouth," the youth said later. There was the adulation, but as always there was also the hate.

Kennedy again returned to Washington that night for more campaign planning at Hickory Hill. The next morning, his young Senate aides briefed him for an appearance on *Meet the Press,* at NBC's Washington studio, but he did not come off well. Four prominent political reporters—David Broder of the Washington *Post,* columnist Robert Novak, Tom Wicker of the New York *Times* and Sander Vanocur of NBC News—tore him apart on the very point he so wished to avoid—his "ruthless" entry into the race. He had held back, he had told himself and others over and over again, because he did not want his campaign to be reduced to a personality struggle between him and President Johnson. But getting in after McCarthy's strong New Hampshire showing made him seem even more ruthless.

All the panelists had a go at the guest, chiding him on his intent to "cooperate" with McCarthy by noting that he was competing with the Minnesotan in key primaries. Would he step aside in California, Broder asked, if McCarthy beat him in Nebraska and Oregon? "I think we can work that out," Kennedy replied weakly. And Novak inquired: "Senator Kennedy, quite apart from your own candidacy, which, you said, if you had gotten in earlier would have created a personality clash with the President . . . if you feel this deeply about the issues, why didn't you support Senator McCarthy in the New Hampshire primary? That wouldn't have been a Kennedy versus Johnson fight."

It would have fed the charges, Kennedy argued plaintively, "that this was a conspiracy on the part of the two of us, and that he was just in there as a front for me, and I don't think that therefore we would have made a contribution on the issues by my taking that step." But he wasn't very persuasive.

From *Meet the Press,* the Kennedy campaign—still in a grand state of disorganization—boarded another regularly scheduled plane to Boston. There, despite the poor weather, the Irish of South Boston turned out for their St. Patrick's Day parade with a much warmer and more forgiving attitude than the candidate had met in New York the previous day. Together the Kennedy brothers strode along the route, two Boston boys back home on the special day, and if he was a ruthless opportunist, he was their own. Kennedy and his aides talked on the flight to New York that afternoon about the difference between the Boston and New York receptions, and somebody said that to the list of things politicians should avoid like the plague—racetracks and sporting events—New York's St. Patrick's Day parade certainly should be added.

By the time Kennedy got back to New York, the news of the secret Vietnam commission negotiations wasn't secret anymore. Roger Mudd of CBS learned about it and beat the news magazines to the punch. Asked about it at the airport on his arrival, Kennedy at first said he would have no comment. But when he found the television crews waiting at his United Nations Plaza apartment he knew he would have to reply. The White House was attacking and he would have to counter. He picked up the phone and called Dick Daley in Chicago, who according to the Kennedy version had been one of the originators of the commission scheme. The candidate told the mayor the only reason he had discussed the plan with the Administration was because he thought it was a legitimate idea and a way to avert the need for him to run. "I'm shocked," he told Daley. "You know all about this. What should I do?" Kennedy didn't say so, but it was obvious he hoped Daley himself would speak out. Instead, the advice he got from this experienced and professional politician astounded him even more than the news of the White House leak. Daley told him: "Deny it." Amazed that the mayor could be so naïve about "media politics" or so bullheaded, Kennedy hung up the phone and gathered his aides around to prepare a reply.

Salinger, Mankiewicz and vanden Heuvel worked over several drafts, with Kennedy penciling in additions and corrections. The statement gave the Kennedy version presented here earlier: that Sorensen suggested the commission, not knowing of Daley's interest (he still was identified only as "a Democratic political leader who was a friend of mine"); that Johnson asked Sorensen to offer names; that they met with Clark Clifford at the Pentagon and suggested commission members (but did not insist on Kennedy among them); that the idea was flatly rejected by the President, leaving Kennedy with no recourse but to run.

Kennedy was to go to Kansas on a late afternoon commercial flight for two long-scheduled campus appearances. The flight was held as patient reporters accustomed to such delays—and regular passengers variously irate or excited about waiting for a Presidential candidate—sat and studied the rainy weather outside the LaGuardia terminal. Finally he appeared, New York's own brand of man-eating photographers and television interviewers crowding around him, barking questions. He was calm and cheerful, shaking hands with newsmen who had just joined the traveling party, moving down the carpeted chute into the awaiting jet, greeting staff members and other passengers who still were coping with the outlandish reality that they suddenly were part of a Presidential campaign.

After dinner Kennedy, in shirtsleeves rolled above the elbow and tie loosened, held court for the traveling reporters. Perched on the arm of his chair next to the window, he talked on a background basis—not directly attributable to him—about the Vietnam commission. First of all it was supposed to be confidential, he said, tapping on his teeth with his fingernails, scratching his arm. He told of his own early meeting with Daley—whom he still identified only as "a Democratic political leader who was a friend of mine and supporting President Johnson"—at which the Chicago mayor first suggested the commission with Kennedy as chairman. At that time, Kennedy said, he told Daley it would be better if somebody Johnson wanted headed the group. When Sorensen went to the White House the previous Monday, he went on, he suggested the commission in response to an invitation from Johnson for ideas—without knowing of Daley's interest and without any suggestion or authority from Kennedy. The burden of Kennedy's explanation was that, far from trying to blackmail the President, the whole thing had come about from a Johnson initiative and a desire by Kennedy to stay out of the Presidential race if something could be done about the war. "I didn't want to run for President," he said several times.

"But when he [Johnson] made it clear the war would go on, that nothing was going to change, I had no choice."

It was dark at the airport in Kansas City, Missouri, when the flight landed, bringing Kennedy to his first campaign stop in the nation's great heartland. His first two days of exposure, in Washington, New York and Boston, had been more of a warm-up than a test. He had been in his own backyard, in areas of strong anti-Vietnam and anti-Johnson sentiment, and the enthusiasm for him was to be expected. What happened at Kansas City, however, was not.

Floodlights bathed the plane as it taxied to the terminal, and a wildly cheering crowd surged to the bottom of the plane's ramp, waving signs that said, WE WANT BOBBY and ROBERT KENNEDY FOR PRESIDENT. The Kansas City stop had not been scheduled as a rally, but as a transfer point. Kansas' Democratic Governor Robert Docking's private plane was waiting for the Senator and Ethel, who had met him there. But by happenstance or good surreptitious advance work, a turnout worthy of a general-election campaign for the Presidency was on hand. Shrieks and applause filled the night air as he walked down the ramp, characteristically brushing his tousled forelock back with one hand. From a lower step, he spoke briefly to them, his strident, urgent tones all but lost in the frenzy. At the end, he raised his fist, shook it at them, pounded it into his hand and shouted, "We're going to win!" There were more cheers, and the crowd immediately pulled him and Ethel to it. He went forward willingly, reaching out to take hands, lost with his wife in a sea of pressing bodies and grinning faces. Local police tried to clear a path for them to Docking's light plane parked about 100 feet to the left. But Kennedy pushed ahead to the terminal, his face set in a determined half-smile, half-grimace, pumping hands, taking some and just touching others with his fingertips. Jerry Bruno, looking like one of those bold and impertinent Walt Disney mice, burrowed in to become point man of the official party, warding off bodies, barking orders to police twice his size, his mind only on the safety of his candidate. Finally they turned and worked their way to the small blue-and-white Cessna. Ethel, resplendent in a pale green coat, climbed in, her husband after her. He hit his head on the overhanging hatch, ducked, turned and waved, then disappeared from view.

At Topeka, it was more, considerably more, of the same. "It sure beats those Army bases," grinned Mankiewicz, alluding to Johnson's captive trips "into the country" of those days. An estimated 2,000 citizens lined the fence, and Kennedy worked it, walking slowly, shaking hand over hand, looking straight into ecstatic faces, murmuring "Thank you, thank you very much," and little else. Ethel strolled behind, often coming up to backs of heads, heads still following the familiar profile that had moved by, bringing with it that painful wrench from the past.

He did not, really, look that much like the late President Kennedy when you thought about it—shorter, thinner, less handsome, a bigger and more prominent nose, much toothier, less confident in manner, more casual in dress. But still it was there, born not simply of family resemblance but of that past pain, of television images etched in remembrance—at the shattered widow's side as the body came back from Dallas and was lowered in a nation's grief from Air Force One; with her in the Senate rotunda as the thousands who had waited hours in that longest night moved by the casket; with her as she stepped with forced majesty down the White House steps to lead the cortege of world notables up Pennsylvania Avenue to St. Matthew's Cathedral; with her, finally, on that Arlington hillside when she accepted the folded flag and turned, stumbling but guided by his strong hand away from the last punishing scene.

Robert Kennedy knew this feeling was there, of course, and was touched by it. He liked to think he was carrying forth the torch of him who had fallen, but at the same time he was sensitive to the charges that he was using the past to insure his own future. Many said he milked the legend, and in these first frenetic, emotion-filled nights and days of the new campaign, he helped them make the case. As far back as 1966, when he campaigned across the country for Democratic members of Congress who had supported his brother's New Frontier, he had displayed a baffling ambivalence toward what had been. It was irreverent for anyone to refer to the late President as "Jack" in his presence; to do so would bring an instantaneous freeze. Yet he had not the slightest reluctance then to invoke the image, name and staccato speaking style of his brother on the stump, to arouse the emotions and, unwittingly or not, the memories of the thousands who heard him. It caused one columnist close to him to suggest one night in 1966 that Robert Kennedy really did not know who he was, and who he was supposed to be: the extension of his brother, a living memorial to him, or something else. In the next years he lived with this search for identity, openly working for the fulfillment of John Kennedy's ideas, yet striving always to project a motivation that was grounded not in bitterness toward his brother's successor, but rather in policy positions he himself had labored to refine. It was, no doubt, why the "vendetta" charge plagued him so. He had no intention of ceasing to be his brother's brother, nor could he be. The question was, could he be more than that, on his own; and—finally—how, with all the adulation steeped in the emotional and tragic past, would he know if he succeeded?

Kennedy, his gold cufflinks ripped away by now, mounted an airline ramp with bullhorn in hand. "I come to ask for your help," he said in the familiar Yankee accent. "We have a hard five months ahead and the odds are heavily against us. . . . I run just basically because I think the United States can do better. . . . We don't need to accept the divisions between races, between age groups, the divisions over the war in Vietnam. I need your help, I need your assistance. . . . I ask you to go out to your towns, your villages, your cities, factories, places of work . . . it can make a difference in the direction the United States is going, and I come here to ask you to help. . . . We can change the direction of the United States. We can make the United States stand for something here and stand for something around the world. Give me your help! Give me your help!"

Strident, repetitive, pleading; each time, as he pounded his fist or reached his hand out in awkward supplication, the crowd cheered and somebody would yell: "You've got it!" They had been shouting "We want Bobby!" for two years now, but suddenly it hit them that they could have him—not in 1972, but now. And even here, in heartland Kansas, where love of God and Country would have been thought to cast this war critic as a national villain, they said they wanted him.

Stan Mitchell, a broad-chested farmer who shook the candidate's hand at the airport fence, was asked how he felt about the way Kennedy had entered the race, on McCarthy's heels and all that. "I don't care how he got in," Mitchell said. "Just so he got in."

In Topeka that night, as everywhere else, it was the end of St. Patrick's Day, and even in Topeka there were the Irish. At the Ramada Inn, they were wearing green and dancing, so the Kennedys from Boston, New York and McLean, Virginia, dropped by and shook hands for an hour or more, as an uncertain tenor sang "Danny Boy" and "Galway Bay." It was a fitting close to the first full day of Robert Kennedy's campaign for the Presidency. The candidate and his wife retired to the home of Governor Docking, buoyed by the Kansas reception but most of all relieved that the uncharacteristic pussyfooting of the last long months had been swept aside; it would be all stops out from then on.

It was in the Ramada Inn in Topeka that "Gladys" made her first of many appearances during the Kennedy campaign. "Gladys" was a major boredom-chaser among the reporters; in every crowd across the land from then on they looked for her, and usually found her—a frumpy, oversized, bleached blonde in her late thirties or early forties, always in a sweater (usually green) too small for her and a skirt or capri pants (usually white) too tight, standing there being "sent" by Bobby. Every town seemed to have its "Gladys," and the bigger ones had a bevy (or a stable) of them. One night, after a typical crowd had been surveyed from the press bus, a reporter cried out in dismay, "Where's Gladys? I don't see her out there." There was silence as eyes searched the throng. Then a voice: "Of course. It's Thursday. Bowling night." In time, the exercise grew until Gladys experts also could spot her husband (usually sitting on a stoop in his undershirt sipping a beer) and her kids (invariably the dirtiest) in every crowd.

In all the days yet to come in the campaign of Robert Kennedy, few were to exceed in emotionalism the next one—Monday, March 18—in the state of Kansas. Of the many crowds that had received him, none ever had been more boisterous and enthusiastic than the nation's college students. In the way he looked and in what he said, he came closest of any of the national political figures then prominent to touching the student generation. His self-deprecating wit, particularly, reached the kids, along with his direct appeal to their best impulses for public service. Although the day's two main stops—Kansas State University in Manhattan and Kansas University in Lawrence—had been scheduled in advance of his candidacy, they were ideally placed for his first formal campaign speeches.

Some 14,500 students and faculty jammed the KSU field house. Kids literally hung from the steel rafters, their feet dangling. At this, his initial confrontation with issue-oriented America, there were a large number of McCarthy signs. GENE FOR INTEGRITY one said, and another, RFK: LEADER OF YOUTH OR REBELLIOUS OPPORTUNIST? And there were anti-Kennedy placards: RFK

PROLONGS THE WAR, FATHER HO LOVES BOBBY. But overwhelmingly they were for him: I LOVE BOBBY, KISS ME, BOBBY, BOBBY IS GROOVY, BOBBY IS SEXY, SOCK IT TO 'EM, BOBBY. If there was a definite adolescent tinge to the slogans and to the enthusiasm, it did not seem to matter just then. The overall impression was one of surging public demand, and that was the impression the candidate needed to get his message to the party power brokers, and—one suspected— needed for himself.

At KSU, Kennedy delivered his opening campaign blast on Vietnam. But first he softened up the audience with wit reminiscent of his brother. When Senator James Pearson, a Republican and friend, gave him a glowing send-off, Kennedy said: "I appreciate the introduction by Senator Pearson, and if you examine his words carefully, *I* thought it was an endorsement." He confessed that his "reassessment" of his political position had been a mistake. "Yesterday there was a man from the Internal Revenue Service out reassessing my home." The stories about the proposed Vietnam commission really exaggerated the differences between himself and Johnson. "The only difference was the makeup. I wanted Senators Mansfield, Fulbright and Morse, and the President, in his own inimitable way, he wanted General Westmoreland, John Wayne and Martha Raye."

The crowd laughed. That was to be expected. But when he launched into his text—an aggressive, militant, free-swinging assault on the President's Vietnam policy—the reaction was not. It was thunderous, in heartland Manhattan, Kansas. On a college campus, yes, but at a university hardly considered a bastion of the new politics and student revolt.

"If in this year of choice we fashion new politics out of old illusions," Kennedy said, "we insure for ourselves nothing but crisis for the future—and we bequeath to our children the bitter harvest of those crises. For with all we have done, with all our immense power and richness, our problems seem to grow not less, but greater. We are in a time of unprecedented turbulence, of danger and questioning. It is at its root a question of the national soul. The President calls it 'restlessness'; while Cabinet officers and commentators tell us that America is deep in a malaise of the spirit—discouraging initiative, paralyzing will and action, dividing Americans from one another by their age, their views, and the color of their skins."

It was at that national soul that Kennedy intended to take dead aim, and of all the ills that beset it, Vietnam was in his mind the greatest, the prime contributor to the rest. He did not want, he assured them (for this was Kansas, and how was he to know the reaction), "to sell out America's interests, to simply withdraw, to raise the white flag of surrender [Applause]. . . . But I am concerned . . . that the course we are following at the present time is deeply wrong. . . . I am concerned that, at the end of it all, there will only be more Americans killed, more of our treasure spilled out, and because of the bitterness and hatred on every side of this war, more hundreds of thousands of Vietnamese slaughtered; so that they may say, as Tacitus said of Rome: 'They made a desert and called it peace.' I don't think that's satisfactory for the United States of America. [Applause] I do not think that is what the American spirit is really about. I do not think that is what this country stands for." (More applause.)

Kennedy, as he was to do throughout the campaign, took note of his own early role in shaping Vietnam policy in his brother's administration, saying he would have to bear his share of responsibility. "But past error is no excuse for its own perpetuation," he said. The recent report of requests for 206,000 more troops was just another step in the national self-delusion; the battle for the hearts and minds of the Vietnamese people was being lost; corruption riddled the Saigon government.

"The facts are that eighteen-year-old South Vietnamese are still not being drafted; though now, as many times in the past, we are assured that this will happen very, very soon. The facts are that thousands of young South Vietnamese buy their deferments from military service while American Marines die at Khe Sanh." And departing from his text, pounding the rostrum with his fist, voice rising, he added:

"I don't find that acceptable. If the South Vietnamese government feels Khe Sanh is so important, let them put South Vietnamese soldiers in there and let them take the American soldiers out!"

The crowd roared its agreement, and for a second Kennedy stood there, his fist still clenched, like some courtyard revolutionary ready to lead the aroused out into the streets. It was not, of course, that the South Vietnamese government felt that the Khe Sanh oupost was so important. The decision to defend it so stubbornly and at so high a cost was an American one. Kennedy's words also overlooked the fact there was a South Vietnamese ranger battalion at Khe Sanh and other Saigon units in reserve nearby. But the crowd did not know this, or for the exhilarating moment, did not care.

Kennedy went on, playing the audience with a skillful change of pace. Hammering hard on the war, then easing up with a light touch, then back to the attack: "President Johnson has responded to the criticism of corruption in Vietnam by reminding us that there is stealing in Beaumont, Texas. I for one do not believe that Beaumont is so corrupt. If President Johnson is not going to stand up for Beaumont, Texas, I'm going to stand up for Beaumont. I deeply resent the attack on Beaumont. . . ." (Laughter.)

He recalled the words of the American commander in the town of Ben Tre who said that it had become "necessary to destroy the town in order to save it," and asked why the recapture and destruction of Ben Tre and Hue "should fall to American troops in the first place." Of the destruction, he added, again aiming at the national soul: "I'm responsible and you're responsible because this action is taken in our name. . . . We must ask our government, we must ask ourselves: where does such logic end? If it becomes 'necessary' to destroy all of South Vietnam in order to 'save' it, will we here in the United States do that too? Is that what we want? And if we care so little about South Vietnam that we are willing to see the land destroyed and its people dead, then why are we there in the first place?" (Heavy applause.)

All wars take their toll, Kennedy conceded, "but the costs of the war's present course far outweigh anything we can reasonably hope to gain by it, for ourselves or for the people of Vietnam. It must be ended, and in my judgment it can be ended, in a peace of brave men who have fought each other with a terrible fury, each believing he and he alone was in the right. We have prayed to different gods, and the prayers of neither have been answered fully. Now, while there is still time for some of them to be partially answered, now is the time to stop." And he repeated his formula: negotiate with the NLF; assure it a place in South Vietnam's political life; begin to de-escalate the fighting; insist that the Saigon government broaden its base, institute real reforms and deal with the NLF.

The young audience by now was in his hands, and he made the most of it: "So I come here today, to this great university, to ask your help; not for me, but for your country and for the people of Vietnam. [Thunderous applause.] You are the people, as President Kennedy said, who have 'the least ties to the present and the greatest stake in the future.' I urge you to learn the harsh facts that lurk behind the mask of official illusion with which we have concealed our true circumstances, even from ourselves.

"Our country is in danger: not just from foreign enemies; but above all, from our own misguided policies—and what they can do to the nation that Thomas Jefferson once said was the last, best hope of man. There is a contest on, not for the rule of America, but for the heart of America. In these next eight months, we are going to decide what this country will stand for—and what kind of men we are. So I ask for your help, in the cities and homes of this state, in the towns and farms; contributing your concern and action, warning of the danger of what we are doing—and the promise of what we can do in the future. *I ask your help!* I ask you, as tens of thousands of young men and women are doing all over this land, to organize yourselves and then to go forth and work for new policies"—departing from his text, Kennedy looked up and around the huge field house, reached out with upturned hand, and drove the message home—"not just in Southeast Asia, but here at home as well, so that we have a new birth for this country; a new light to guide us. And I pledge to you, if you will give me your help, if you will give me your hand, I will work for you and we will have a New America!"

The place erupted in a stampede of cheering and applause. Kennedy, seemingly spent, subsided as the roar went on. He had the crowd with him, and its excessive adulation and response—as subsequent events revealed—had him.

After a mostly playful question-and-answer period (Q: "Put yourself in President Johnson's place . . . A: "That's what I'm trying to do.") Kennedy moved out through the pawing mob for the drive to Kansas University at Lawrence. There, an even larger (17,000) and more enthusiastic crowd awaited him in the Phog Allen Fieldhouse. It was the greatest political gathering in the history of the campus, rivaling even the outpourings that came to see Wilt Chamberlain lead Kansas to national basketball prominence in the late fifties. Students stamped and clapped in rhythm as the candidate entered; an overflow crowd sitting on the shiny wood floor got up and swirled around him, leaving only a small island of officialdom waiting on folded chairs in the center. From high above, in the press section, it looked and sounded like some overly done scene from a Hollywood movie of a Presidential campaign—the jumping young girls, the screams ricocheting off the distant field house walls.

Here, Kennedy used the same combination of self-deprecating wit and fierce, emotional opposition to the war, and it had the same lightning effect. After Pearson, his Senate colleague, had been introduced, Kennedy said: "I was sick last year and I received a message from the Senate. It said, 'We hope you recover.' The vote for sending it was forty-two to forty." And he told of a poll of businessmen on their preference for President: "I received one vote, and I think they're looking for him."

Then the candidate launched into the Vietnam speech again, this time off the cuff. But all the points were the same. And when he spoke of corruption in Saigon and draft dodging by South Vietnamese youth, he threw all the indignation he could muster behind the line about American Marines having to defend Khe Sanh. The reaction again was thunderous, and those who had heard the earlier KSU speech could see how he was using the emotional response to build more and more of it; how he himself seemed to be pulled up on it like a small boy on a towering seaside breaker, riding it willingly, daringly, with evident exhilaration.

Deftly, he moved from troubles abroad to troubles at home, and the crowd with him. "We have seen tanks patrolling American streets, machine guns fired at American children. I don't think that is a satisfactory situation, and that is why I run for the Presidency of the United States."

When it was all over, and Kennedy had bulldozed his way through the cheering mob and was heading back toward Kansas City, the question of a severe negative reaction to the manner of his entry into the race seemed to have been swept away by sheer adulation. True, the settings had been campuses, where Kennedy always had been popular; also, at least half—probably more—of those who heard him at the two universities could not vote for him, and certainly would have little direct impact on the nominee-selecting process within the Democratic Party. But there seemed to be little doubt about the depth of antiwar sentiment, or anti-Johnson sentiment, and that was Kennedy's biggest weapon.

On the way back to the airport for the trip home to Washington, there occurred one of those little-reported but unforgettable episodes that can blur the impact and the effort of a singularly effective campaign day, for those who have seen it. When it happened, it was raining hard outside, and not all the reporters in the press bus got off. The main events of a long and wearing day were over, and deadlines in the East were pressing.

Kennedy, as a member of a Senate subcommittee concerned with American Indians, had become particularly interested in them, and in his speeches he often had dwelt on their plight, quoting statistics about their staggering suicide rate. And so, when Mankiewicz announced on the press bus that an unscheduled stop would be made at nearby Haskell Institute, "a terminal trade school for Indians," immediate gag-making was in order. Somebody asked what a terminal trade school was. With the suicide rate in mind, Jimmy Breslin offered: "When they graduate, they hang themselves."

In the small auditorium of the school, the students greeted Kennedy with the usual enthusiasm. They were a bright-looking lot, predictably so because they came from reservations all over the country—select students. Kennedy asked how many came from this state and that state, and the hands went up accordingly. He introduced Ethel and Rene Carpenter, wife of Astronaut Scott Carpenter, and proceeded to make the most obvious and blatant pitch for the Indian vote—though there could not have been many votes in the hall. He discussed only "the Indian problem," of which he said patronizingly, "This is not an Indian problem. This is a white man's problem," and he went on as though he had just descended on a reservation where the inhabitants never had heard of Vietnam or any of the other general problems tearing at the country. He continued, sometimes being embarrassingly flip as the young students sat politely, and topped off the performance by saying to them: "I hope if no Indian is running for President of the United States, you will vote for me as the next best thing." The applause was moderate, as if the audience was not at all certain of that. Then Kennedy was off on his way to Kansas City. He had not had to make the stop; he did not have any time to plan what to say and very little time in which to say it; he had had an extremely busy and emotional day. His public record toward American Indians certainly was above reproof—in fact, he probably was one of their most diligent, outspoken champions in Congress. But he also was a candidate for President now, and it was one of those deplorable lapses into which candidates fall. Fortunately for Kennedy, not many of the press had seen it, and some who did were not bothered by the performance or had more dramatic things to write about that day.

On the flight back to Washington, Kennedy was relaxed. "How do you feel now that you're in it?" a reporter asked him. "I feel good," he said, and it was obvious he did. "All those things I was saying before [about supporting Johnson]—that wasn't me. The others, those who didn't want me to run, they never got out in the country and saw the crowds. I saw them in 1966. They didn't know what I knew. I felt it in my gut. I had to go." And now he was going, full-tilt.

But before Kennedy shifted into high gear, there was one other matter to attend to. When one Kennedy went, they usually all went. This time, however, there was a complication. Sargent Shriver, Eunice's husband, still was playing on the other team. His decision to stay on with the Johnson Administration always had been a source of consternation in the family, as was his brief flirtation with the Vice Presidential nomination in 1964 and his conduct as head of the antipoverty program. He was not the favorite brother-in-law; Robert Kennedy once had referred to Sarge as having "a Kiwanis mind." But for family reasons, the candidate decided to ask him to come aboard.

During the critical decisions on his candidacy, Shriver was in Europe, preparing to take a new offer from Johnson to be the American Ambassador to France. The brothers-in-law talked by phone, and again when Shriver returned to his home in Rockville, Maryland. Finally a breakfast meeting was arranged at Hickory Hill, at which Kennedy invited him straight out to join the campaign. Sarge said he would have to think it over. Later in the day, he called and left a message: he was going to Paris. Apparently no involved reason was given. Those least favorably disposed to Shriver inevitably assumed he had weighed his brother-in-law's chances against the prospects of Johnson's reelection, and had acted accordingly. Others saw in the decision Shriver's determination to be on his own, free of family strains that had built up over the past years. In any event, his rejection of Kennedy's offer, needless to say, did not further endear him in the family ranks. Later on, in Indiana and elsewhere, Eunice Shriver campaigned for her brother—a role not usually assumed by an ambassador's wife. It was surprising the McCarthy forces never specifically challenged it.

The "blow" of Shriver's loss to the Kennedy campaign was accompanied by a public development on March 21 that suggested more clearly than before that Kennedy's fears about the near-inevitability of Nixon's nomination by the Republicans were well-founded. Nelson Rockefeller, after revving up his liberal and moderate supporters with mounting indications that he would enter the field against Nixon, abruptly announced that he would not. Rockefeller told an astonished nationwide television audience that extensive party soundings had convinced him the party didn't want a divisive contest against Nixon. He therefore would not enter any primaries or become an active candidate. From the Kennedy viewpoint, Rockefeller's nomination by the Republicans would have increased his own chances to become the Democratic nominee, but if others had expected that to happen, Kennedy had no illusions on that score.

Kennedy's next trip, three days after the Kansas swing, was a one-day overture into the South, again to two college campuses, to provide the base for later Kennedy claims that he had carried his campaign to all parts of the nation. The excursion into Georgia, Alabama and Tennessee was approached with some trepidation by staff and friends; deep anti-Kennedy sentiment said to exist in the South was accepted as fact by most. But in stops in Atlanta, at the University of Alabama in Tuscaloosa and at Vanderbilt University in Nashville, Kennedy again drew large and friendly audiences. His devil—Lyndon B. Johnson—was their devil. He made certain at each stop to emphasize he was not advocating unilateral withdrawal from Vietnam, and at Vanderbilt he decried "anarchists [who] threaten to burn the country down." But having said that, he tore into the Johnson Administration—not by name, but by the clearest implication. Defending the role of dissent and his own conduct as a dissenter, he told the students:

"When we are told to forgo all dissent and division, we must ask: who is it that is truly dividing the country? It is not those who call for change, it is those who make present policy . . . who have removed themselves from the American tradition, from the enduring and generous impulses that are the soul of this nation." He warned that "every dictatorship has ultimately strangled in the web of repression it wove for its people," and that as a result of divisions in the land young people were turning away from their "public commitment of a few years ago to lives of disengagement and despair, turning on with drugs and turning off America." The strong words led Richard Harwood of the Washington *Post* to report:

"Sen. Robert F. Kennedy invaded the South today and accused President Johnson of creating the worst divisions in America since the Civil War. Going further, he implied that the President is to blame for the alienation and drug addiction among American youth, for rebelliousness and draft resistance on American campuses, and for the 'anarchists' and rioters in American cities."

The indictment was pretty strong, but it reflected a spreading uneasiness among the campaign press corps about the free-swinging oratory of this young and intense politician whose words and presence seemed to have such an emotional impact on those who heard and saw him. One reporter who had been in Kansas and the South confided to a colleague who had been on neither trip that at times it had been "scary." He could see the audience respond, he said, and then could sense that the fervor created was having a playback reaction on the speaker, leading him to escalate the message and the mood.

In the next several days, similar remarks were heard increasingly as the Kennedy campaign moved across the country on its first major swing—a nine-day trip to the West Coast, the Rockies, the Midwest, the Southwest and home. It was the real shakedown of the campaign. Although the chief traveling aides like Mankiewicz, Dutton and Dick Tuck, the Peter Sellers of politics, knew they had a charismatic candidate on their hands, they may not have fully grasped his potential for stirring the populace. Simply his presence—the Kennedy mystique and all that, the emotional ties to the past—constituted a tremendous pulling force. But as the personification of a nation that was bitterly, even unreasonably, distrustful of its President, and as a man now all at once freed of a long self-discipline against open criticism of Johnson, Robert Kennedy became a force on the stump seldom seen in American national politics.

The entourage, a gay mix of politicians, social hangers-on and newsmen, boarded a regular commercial flight at John F. Kennedy International Airport in New York on the bleak, misty Saturday morning of March 23. The candidate, who Friday night had been in upstate Binghamton for a testimonial dinner for John Burns, the New York State Democratic chairman, had put in an exhausting night getting back, and as soon as the plane lifted off, he threw a pile of pillows and a blue blanket on the floor and tried to get some sleep, as photographers recorded the scene for posterity. Among the notables aboard was Jesse Unruh, who had the latest favorable polls in hand and was available to reporters who wanted an interpretation. They showed, generally, that Kennedy would beat Johnson handily in a two-man race, and even would run ahead in a three-man race, including McCarthy. Johnson would not be on the ballot; State Attorney General Thomas C. Lynch, the highest-ranking officeholder the Democratic orthodoxy could get, would head a slate for him. The California Democratic Council, an independent left-of-center powerhouse of more than 30,000 members, was the backbone of the McCarthy effort, and for the time being was holding firm. But the Lynch slate, Unruh and other Californians told the press, already was falling apart. Wives and sisters of Lynch delegates were coming out for Kennedy; it would be, the newsmen were assured confidentially, only a matter of time before the pro-Johnson slate would be reduced to ineffectiveness.

The jet arrived in San Francisco with a fairly well-rested candidate, although his floor nap did not last very long. Pierre Salinger, who had done it all before, bounded aboard with latest details for the next two days of whirlwind campaigning. He, and the other old, recognizable faces from 1960, had an assurance of success about them that was intimidating. In the terminal, the crush was predictably impressive. The tell-tale indicators of solid advance work—the similarity of signs, of campaign hats and stickers—were all there, but the fervor of the crowd suggested how easy it was to draw a crowd for Robert Kennedy. The pockets of dissent were there too, as always. A sign said WELCOME, BOBBY-COME-LATELY, and its holder, a blond matron, shouted: "Where were you in New Hampshire?" To which a woman next to her snarled: "New Hampshire is nothing. Look around you." And she gestured at the straining, jumping multitudes.

Two Boeing 727 charter jets were waiting to take the party to the first stop, Stockton, to the southeast, and on to Sacramento, San Jose, Salinas-Monterey and Los Angeles in the next day and a half. From then until midday Monday, when the party left for Oregon, it was a mad blur of landings, motorcades, mobs, plaintive speeches, more motorcades, more mobs, takeoffs and the whole cycle over again. Kennedy had gone to the campuses for issue-oriented audiences, and they had responded with raw physical emotion. Now he went into the streets, where that reaction was to be expected, and it materialized in an intensity and scope that was awesome and frightening. Everywhere, Kennedy delivered the hard-hitting Vietnam speech, bearing down on the lines that had produced the wildest response on the campuses, and the result always was the same. The familiar Kennedy profile and gestures, the rising New England voice, the heavy repetition of phrase, the unruly forelock and the hand absently brushing it back in midsentence—all these invoked memories of his brother, and he did nothing to discourage that reaction with phrases such as "this is a time to begin again, and that is why I ask your help, and that is why I run for President." And the crowd, for its part, contributed to the spirit of New Frontier resurrected. At the Sacramento Airport, a sign read: CALL THE ROLL AT THE ROUNDTABLE. CAMELOT WILL COME AGAIN. ROBERT KENNEDY WILL BE PRESIDENT AGAIN (sic). And at a shopping center in suburban Sacramento, another said: CONTINUE THE NEW FRONTIER.

In that shopping center, in midafternoon of Saturday, there occurred a mob scene that in size, frenzy and the physical threat of stampede rivaled anything that DeMille ever had committed to film. Kennedy spoke from a ladder in a mall inside the huge new shopping plaza, with the crowd packed around him on all sides, extending well back up the aisles of the department stores that opened onto the mall. The crowd was boisterous and rambunctious, but he played with it and brought it under control with an emotional plea for an end to the war. "Which of these brave young men dying in the rice paddies of Vietnam might have written a symphony?" he asked, his voice dropping. "Which of them might have written a beautiful poem or might have cured cancer? Which of them might have played in a World Series or given us the gift of laughter from a stage, or helped build a bridge or a university? Which of them might have taught a small child to read? It is our responsibility to let those men live."

When he had finished, and began to move out of the courtyard, the crowd, which had greeted his soliloquy with sustained applause, suddenly became a live and dangerous thing. It surged toward him from all sides, pressing onlookers against store windows until the glass threatened to buckle; children were being swept along and pushed underfoot. Kennedy aides, trying to clear a path for the candidate, had to reach down and rescue children who had slipped.

When Kennedy finally made his way to his open convertible and climbed aboard the trunk, he was practically pulled off. Bill Barry, a former Kent State football star, FBI agent and now bank security official on leave to be the candidate's personal security man, knelt next to Kennedy, locking his strong left arm around his waist to keep him from being pulled over.

It was a scene that was to be repeated for thousands of miles of motorcades across the breadth of the United States in the next ten weeks. On one night, Barry knelt for most of nine hours through the ordeal, until his knees were rubbed raw and bleeding. Ethel Kennedy got him a rubber kneeling pad, and that helped; but still it was sheer physical punishment. Yet Barry, perhaps more devoted to the candidate than any other member of the party, maintained his constant cheerfulness and gentleness.

For all his past training, Barry was not prepared for the size and aggressiveness of the crowds. As an FBI agent in New York, he had served as a security escort for Kennedy in his 1964 Senate campaign and later in the Senator's frequent political trips around the state. But a national campaign was something else again. "When I got off the plane at San Francisco and saw that first one, the shock sobered me on what the job was," he recalled later. "People hit me, tried to pull me away. At the end of the day, his [Kennedy's] hands and mine would be bleeding. Once, at the beginning, I told the Senator, 'I wish these people would be more courteous.' He answered, 'They're here because they care for us and want to show us.' After that, I never had any trouble adjusting to crowds. I found they wanted not just to touch a celebrity; they wanted to convey their feelings to him, and he accepted it for that. Sometimes when the going got rough in a tight situation, with all the shoving around we took, he would turn to me and say, 'Don't hurt anybody. Don't push so hard.' " There never was any doubt in the mind of anyone who saw Barry at work that his was a labor of love. Much later, he put the obvious into words: "I loved him intensely as a human being, and for his qualities. I was dedicated to see him through this. I wanted him to be President of the United States for the sake of my children and generations to come. It was not just a professional job with me. It was something my life qualified me for. This would be my juggler's gift. This is what I had to offer, and I wanted to give it." Throughout the swift, grueling campaign, no one gave more.

At the Salinas-Monterey Peninsula Airport on Sunday, a picket fence held back a particularly dangerous crush. Kennedy himself became so concerned as children below him were being smothered by mindless adults that at one point he raised his hands and said: "Ssssh. Just be quiet for a minute. Just clear a path for the children." And he reached down and plucked a small girl from the crowd, holding her until the path was opened. He watched silently as the children were led out of danger, and then resumed shaking hands.

Back on the plane, Kennedy expressed concern about the kids, and asked why their parents brought them. But he encouraged this meeting the people, this touching; he did it with a kind of set determination, a resignation that it was all part of what he was and what he represented to them, in the past and in the present. He plunged into these mobs the way he plunged into icy ocean breakers. Nobody forced him, and he never hesitated, yet one always had the feeling that he had to steel himself for it; meeting people was never his strong suit, certainly not in small groups. At least with the mobs, none of the painful, embarrassing small talk was required; he was, in a sense, a national shrine, to be seen and touched so that in later years the visitation could be recalled, or so that small children, on growing up, could be told they had done it.

On Los Angeles' frenetic freeways, it was the same. Motorists would speed by the motorcade, the driver would take both hands off the wheel to snap a picture for posterity, then race on. One motorcyclist, trying to steer with one hand at a breakneck pace and focus his camera with the other, nearly crashed into Kennedy's car. Jerry Bruno reached out, took the camera, made the picture of Kennedy and handed it back to the man on the motorcycle.

In his short speeches at every stop, Kennedy scored heavily with his remarks about Vietnam corruption. Now, when he talked about the draft in South Vietnam, he spoke of eighteen- and nineteen-year-olds not being called. In fact, nineteen-year-olds were being drafted and the eighteens were about to be. Newsmen who knew this, and had passed the information on to Kennedy aides, shook their heads in dismay—and wondered aloud what was being set loose here, in this excess of zeal and yearning and adulation.

The climax occurred early Sunday night, after a frightening reception at the Los Angeles Airport in which thousands pursued the candidate through a long tunnel to the street, their chants of "We want Bobby! We want Bobby!"

thundering deafeningly in his ears, and where a child was pulled from off the rear fender of a motorcade car, bruised, crying, but not seriously hurt. A photographer for a California paper called his office after Kennedy left the airport and shouted into the phone: "They damned near killed him today."

The candidate spoke in the Greek Theater at Los Angeles' Griffith Park, an amphitheater cut out of a hillside. The traffic jam getting there was unbelievable; cars abandoned everywhere and their occupants trudging up the hill in the late afternoon sun to hear the speech. Kennedy, weary but determined to give his best to the throng of about 8,000, and 3,000 or more listening outside, read the text well; it seemed to be no more than a reworking of his ad lib theme—that the war in Vietnam was a grave mistake whose pursuit was shattering the very principles on which the nation was founded.

But near the end he said something that startled the attentive press corps. "The failure of national purpose" evidenced in the trouble abroad and at home, Kennedy charged, "is not simply the result of bad policies and lack of skill. It flows from the fact that for almost the first time the national leadership *is calling upon the darker impulses of the American spirit*—not, perhaps, deliberately, but through its action and the example it sets—an example where integrity, truth, honor and all the rest seem like words to fill out speeches rather than guiding beliefs. . . ."

It was a stinging indictment, yet more than that a vague and shapeless attack, reminiscent of the witch-hunting era of McCarthyism. The audience didn't seem to mind, but influential members of the press, listening with growing consternation to the litany of careless remarks about the war, seized on the phrase as evidence that in Robert Kennedy, intoxicated by the fantastic crowd response, they might well have a budding demagogue.

In the next days, that ugly and devastating word began to appear in print around the country with a frequency that was dismaying to the Kennedy staff. Robert J. Donovan, the respected Washington bureau chief of the Los Angeles *Times,* wrote the next day that a lesson of the early Kennedy campaigning "is that when a war becomes a flaming political issue, the line between debate and demagoguery becomes a thin one. A candidate can easily be carried across it in the ardor of the fight." And Harwood wrote in the Washington *Post* a few days later that the crowd reaction "has led at times to rhetorical devices" that even among members of Kennedy's own staff "have been regarded as bordering on the demagogic."

It didn't take Kennedy long to get the message. His policy with the press always was to keep the lines open, in the hope of picking up valuable indicators of the campaign's progress and direction. He would walk up and down the plane chatting with reporters, and his aides would do the same. In this atmosphere, the candidate and the staff were aware long before the implications of demagoguery hit print that the press was casting a cold eye on the zealous candidate. After the Kansas and Southern trips, they already had culled this attitude, and Kennedy was under pressure from aides to "cool it." Specific references to Johnson were trimmed, but when the California mobs began to work their passions on the candidate, he was increasingly carried away. Although Robert Kennedy had the reputation of being a cold and blunt individual, he was a much more emotional and reachable man than his brother John had been as a campaigner for high public office. John Kennedy, especially in the late stages of the 1960 campaign, could fire up a crowd, but seldom did the playback get to him. Robert was not so emotionally insulated; that was one of the reasons his compassion for the poor was accepted as genuine by the mass of underprivileged with whom this rich and privileged young man came into contact. But this same emotionalism—and the intentional playing on it, to the edge of demagoguery—could also produce a negative counterreaction among those who, in times of national distress, longed for more placid times. A soft-sell candidate like Gene McCarthy, who actually was assaulting the political status quo much more basically than Kennedy was, nevertheless did not project a very frightening image—except to entrenched party hacks.

At any rate, the "darker impulses" speech was a watershed of sorts in the Kennedy campaign. It helped shake down the staff and the candidate himself. To attract the crowds and then work them up was great fun and a great morale booster, but the game also carried with it the seeds of political disaster. In a sense, the episode—this early period of boundless enthusiasm and attack—was akin to the early months of the John F. Kennedy Administration. Then, as in this instance, a free-wheeling cockiness was brought to earth by a sobering setback—the Bay of Pigs. The Kennedy Administration underwent an administrative housecleaning after that and was much more effective, and much more cautious, for it. In Robert Kennedy's campaign, there was no comparative housecleaning; there was not, from all indications, even a full-fledged meeting to straighten things out. The candidate had swung wildly and had been given a bloody nose for his trouble. He picked himself up and went on, a bit wiser, a bit more cautious and restrained.

Yet every drama must have a villain, and the press sought out one for the "darker impulses" speech. Some reporters fastened on Pete Hamill, the former newspaperman turned free-lance writer who was traveling with the press. Mankiewicz acknowledged under questioning later that Hamill had contributed some "ideas" to the campaign but did not say he had written the Greek Theater speech. Hamill later denied in letters to newspapers that he was its author, but many reporters still believed that he was.

Actually, the speech was written for Kennedy by Dick Goodwin while he still was a member of the McCarthy staff. Goodwin had been involved in all the early meetings of Kennedy insiders when the question of running was under exploration, and it was assumed that anytime Kennedy decided to seek the Presidency, Goodwin would be with him. It hadn't worked out that easily, because when Kennedy seemingly had decided against running, Goodwin, bitterly opposed to the war, had joined McCarthy in New Hampshire. Now that Kennedy was in, he and his closest aides knew they could use Goodwin and were pressuring him to leave McCarthy and come aboard. Goodwin felt he had to stay with McCarthy through the Wisconsin primary on April 2, but he wanted to do something. So he offered to write one speech for Kennedy in the meantime, and confidentially. This was it. The draft was read by several Kennedy aides and by the Senator himself before delivery, and in the rush of a campaign that literally was being planned day by day, the "darker impulses" line never jumped out at anybody as a potential troublemaker. But from then on, the texts were screened more carefully.

Jesse Unruh, on the plane out to California at the start of that tempestuous weekend, had said of the task of winning the nomination: "We can't just sit down at a table and expect to bargain for delegates. We've got to produce a groundswell in the country." Within two days, none could dispute that the Kennedy production line was turning out the goods. But it was clear now, too, that public clamor could be overdone. What was the reaction of all those unseen millions of voters who did not show up for rallies and motorcades, who did not jump and scream, who were not teenagers or college students or young marrieds who were turned on by the groovy Bobby Kennedy? The question was increasingly in the minds of Kennedy aides, and again the need to "cool it" was stressed.

It was not true, as many suspected at the time, that the all-out, sometimes reckless beginning was a result of the "wild impetuous Young Turks"—the young aides who had wanted Kennedy to run from the start—gaining the upper hand from the old-timers. There were some early staff conflicts, but not serious ones. The younger men resented the fact that older hands like Sorensen, Schlesinger and Goodwin had the candidate's ear, but all came to respect each other's abilities and contributions. The younger aides, after all, had been right about the temper of the country before New Hampshire, and they knew how Kennedy could most effectively address himself to the nation's activist youth. If they were not consulted on the hard political decisions, they still were heavily involved in the message of the campaign itself. The intemperate nature of the first few days effort, at any rate, could not be laid to them. Rather, it was more that after the months of gnawing indecision and sail trimming, Kennedy could and did cut loose. Mankiewicz later referred to the early slam-bang mood as "the free-at-last syndrome," and that summed it up quite well. "It may have intoxicated everybody a little," the press secretary admitted.

Whether by design or not, the reception for Kennedy in Oregon and Washington the next day, and in Idaho, Utah, Colorado, Nebraska, New Mexico and Arizona all the next week, was relatively subdued compared to that frenzied California weekend. There was something ominously tame about the crowds in Portland at the airport and in a new downtown shopping plaza called Lloyd's Center on the next day, Monday. At the latter, Kennedy stood on a high step of a winding steel staircase and talked almost conversationally to the large weekday gathering. For the first time in memory of many newsmen, the candidate in looking over the crowd remarked that he saw some familiar people "who worked for Jack." It was that single reference and nothing more; but on all other occasions before and after, in public at least, it always was "President Kennedy" or very occasionally "my brother."

Part of the "cooling it," it appeared to many on this trip and later in the campaign, was to stay away from the pointed references to John Kennedy. But the mystique always was there, in public and in private. On one leg of the journey, Kennedy was standing in the aisle of the chartered jet chatting with two reporters. The Senator, as always, was wearing his PT-109 tieclasp. One of the reporters removed his own tieclip, a strange pointed thing with the sides cut away to show wires inside, and handed it to him. With a smile, the reporter asked Kennedy if he knew what it was. Kennedy studied it and said he didn't. "It's a torpedo," the other said. "I wear it when I travel with you just to show I'm staying honest." The reference, of course, was to the PT-109 clips Kennedy and all the staff wore. It was a very bad joke, much too close to home. Kennedy's face froze. He handed the tieclasp back and walked away.

The campus audiences in Oregon, Washington and elsewhere were warm and enthusiastic, but after California Kennedy was relatively low-keyed on Vietnam. At the University of Portland he used his favorite quotation from Camus to impressive effect: " 'Perhaps we cannot prevent this world from being a world in which children are tortured. But we can reduce the number of tortured children. And if you don't help us, who else in the world can help us?' And that is the question I ask all of you." In referring to inequities between the American and South Vietnamese military drafts, Kennedy was more careful: "They're now announcing they are going to draft nineteen-year-olds and eighteen-year-olds. That's fine. But we've been doing that all along."

In his light remarks, he seemed to try to be less "cute"—that word that came with such maddening frequency from the lips of purring young girls and that caused such chagrin among aides trying to project a mature national figure. And, as ever, there was the ruthlessness charge. He had some standing gags about it, like the one about Teddy Kennedy printing campaign buttons with his own picture on them and Robert telling him it was too late to get into the race because it would look too ruthless. But they were contrived, and he handled the problem best in off-the-cuff remarks. At the University of Washington in Seattle, clearly McCarthy country, he nevertheless drew a packed house with standees. Students sitting behind the overflow couldn't see and they protested loudly. "Those of you who really don't have to stand, sit down," Kennedy said in a mild tone. And then he added, to the crowd's delight, "There I go being ruthless again."

Kennedy was not at his most attractive as a fist-pounding rabble-rouser, anyway. Especially with the kids, he got closest to them, ironically, not when he agreed with their views that everything in the country was going to hell, but when he took an honest stand against them and said so. One of his big differences with them was conscientious objection and the draft. He told them he favored deferment for legitimate conscientious objection, but those who opposed the war on other than religious grounds should be willing to pay the penalty of avoiding the draft as long as other Americans, particularly the poor, were dying in Vietnam. He himself would go if drafted, he told them, but he recognized it was a matter of individual conscience. At San Fernando Valley State College in California, one of his last stops there on the wild weekend, he was booed for this view, then cheered when he told the audience: "I know that's not the popular attitude, but it happens to be the way I feel about it. . . . You say, 'Tell it like it is' and tell you the truth and that's what I intend to do." At Seattle, he lectured the pro-McCarthy crowd about the injustice of student deferment in wartime and in the end was applauded for his straightforwardness, though most seemed still to disagree with him.

Everywhere in the next week, Kennedy's crowds were large, but not as large as in California; warm, but not frenzied as in California; strongly for ending the war, but less emotional about it, except for the University of Nebraska, where the antiwar sentiment was thunderous. And for his own part, Kennedy cooled it.

The candidate used part of the week visiting Indian reservations in the Southwest in connection with his Senate subcommittee work. (After the usual suicide statistics, Breslin to Kennedy: "It's a good thing the rope broke for Jim Thorpe.") He took a few hours off to ski in the Rockies. It was a period of relative quiet. In Salt Lake City on March 28, a crowd of 5,000 was waiting for him when the local police received an anonymous call that a bomb had been placed in the hall. After a search was made and no bomb found, the crowd was told. Most elected to stay. "If the people are in there, I'm going in," Kennedy said when he arrived, and he did. "This is known as opening your campaign with a bang," he told the somewhat nervous audience. "I wouldn't want to go with a nicer bunch of fellows." Nothing happened. And at the University of Arizona, booing students unrolled a huge picture of the state's first Presidential candidate and favorite Air Force Reserve General, Barry Goldwater. Kennedy paused, looked and said: "I was wondering what ever happened to him." And shortly afterward, when a low-flying jet drowned out the voice of a questioner, he interrupted to say: "Would you wait until Barry finishes?"

In the daily rush of travel and meeting appointments, the candidate and his closest advisers were coming up to their first major strategic decision—the Indiana primary on May 7. The deadline for filing was the night of March 28, a Thursday. Kennedy had not mentioned that test in his announcement of candidacy. For one thing, he wanted to avoid for the time being at least the inevitable fact that he would have to run against McCarthy. Thus he said in his announcement that state laws required his name to be on the ballot in Nebraska (where Sorensen called at the last minute to make sure he did get on) and in Oregon, and that he would be entered in California "to make certain" the effort against Johnson "will still be effective in June." What this meant, of course, was that he hoped to have McCarthy knocked out of the picture by then, and would be able to take on Johnson alone there.

In Indiana, there was no such convenient or even halfway plausible dodge. If Kennedy went in there, it would be to dispose of McCarthy as early as possible. Indiana was the first primary still open to Kennedy, and it was a notoriously conservative state. Yet it had elected two progressive Democratic Senators, Vance Hartke and Birch Bayh, and offered a chance to get in an early shot against Johnson. The amiable old governor, Roger Branigin, was expected to run as Johnson's stand-in, backed by what was reputed to be one of the strongest Democratic state organizations in the nation. Gordon St. Angelo, the state chairman, had a large and experienced staff and a built-in treasury: certain state employees were required to pay 2 percent of their salaries into the party coffers. The system produced not only campaign funds but a tight base of party dependables who were not likely to do much bolting when their jobs so obviously were on the line.

In addition, Indiana was an extremely insular state and proud of it. You were a Hoosier, or you weren't, and no candidate was likely to stick out more than the nation's most prominent Yankee, Robert Kennedy. He had encountered the carpetbagger label in so sophisticated a state as New York, and had found it to be a formidable political weapon for the other side. Finally, Indiana was the center of old-line Americanism. The national headquarters of the American Legion was in Indianapolis, and the city was graced—or blighted—by not one but two of the hugest war memorials in the country. Were a Martian to land his flying saucer in Monument Circle, he might well take one look, climb back in and beat a fast retreat. Kennedy, as a vehement critic of the Vietnam War, would be taking a considerable risk making such a state the critical first test of his candidacy.

On the plus side, there were heavily populated Negro areas in the fringe cities across the state line from Chicago. But alongside them also were ethnic blue-collar neighborhoods—in Gary, Hammond and the rest of Lake County— backlash country for George Wallace in the 1964 Democratic Presidential primary, when the former Alabama governor received an astonishing 30 percent of the total Indiana vote. And there was the fact that the Negro population on a statewide basis was small, only about 9 percent.

Still, it was important to have a test against Johnson—and McCarthy—at the earliest possible time, and what the campaign needed was not just a pushover, but a real test whose winning would mean something, would provide real thrust for the Kennedy operation. Crowd adulation was one thing, but it had to be translated into votes to really get the message through to the big-city political powers. And the way things were going, the euphoric Kennedy party was extremely confident that despite the handicaps, he would win.

In a meeting at Hickory Hill, Ted Kennedy and Steve Smith had discussed the Indiana primary with the candidate and they had decided tentatively to go in, depending on a special poll being taken by John Kraft. The poll became available on Wednesday, March 27, and it showed Kennedy running close behind Branigin and ahead of McCarthy. The candidate himself decided to go ahead, "on sheer guts," Sorensen said later. When Kennedy was warned that Indiana would be a gamble, Sorensen said, he replied, "The whole campaign is a gamble."

With the poll results in one day before the deadline, Kennedy announced in Salt Lake City that he would enter. After stops the next day in Denver and Lincoln, Nebraska, the campaign plane rushed east to Indianapolis so the candidate could file in person. He was mobbed at the airport and on the grounds of the statehouse downtown, where he submitted the necessary papers for the primary. An estimated 4,000 screaming backers pressed around him roaring at his fervent campaign appeal, as he pushed his way through and into the Capitol. When he said, "Give me your help," the shout came back at him: "You got it, you got it, man!"

Trying to keep the focus on Johnson, Kennedy told the audience that "I am not here to oppose Governor Branigin. He is in no way responsible for the policies and actions I challenge this year." But one of Branigin's major supporters, the Indianapolis *Star*, obviously saw it otherwise. Although Kennedy's arrival had created an uproar, the morning paper buried the fact deep in the lead story, electing instead to note first that the state's attorney general had come out for Branigin, that McCarthy's state chairman had called Kennedy's entry "cutthroat competition" and that Branigin's two sons had filed for him while the Governor was returning home from Florida. It was just the beginning of a treatment that caused Kennedy finally to publicly label publisher Eugene C. Pulliam's Indianapolis *Star* and *News* (the evening paper) "the worst newspapers in the country"—an honor he previously had reserved for William Loeb's archconservative Manchester, New Hampshire, *Union Leader*.

McCarthy, meanwhile, seemed to be pulling in his horns. In an airport press conference in Oshkosh, Wisconsin, the same day, he sought to minimize the importance of Indiana and his participation in the primary. "I committed myself to go in there before he [Kennedy] made any announcement," McCarthy said, "and at the time all they really said was just give us your name and come in and make one or two appearances. That is what I will do. I will have to conduct a very limited campaign in Indiana. I will probably make three or four stops in the larger cities and do some television and probably some radio and let it go at that." Well, if Kennedy were to win there, wouldn't the Minnesotan be under pressure to get out of the race? "I do not think so," he replied. "It might if I campaigned very thoroughly and made a great effort in Indiana, but I do not see his achievements as having put any particular pressure on me at any point. We will go down all the way and the showdown, of course, will come in California."

After the Indiana filing, Kennedy and party flew to the Southwest for appearances the next two days in New Mexico and Arizona. It had been an exhausting week; at the University of Arizona, he began to stumble over routine words, and at one point in his speech he looked up and said, "God, it's been a long, long day."

Actually, it was hard in those first dizzy two weeks after his declaration of candidacy to recall when one day ended and the next began. In fifteen days of criss-crossing the nation, Kennedy had visited sixteen states with nearly 800 of the 1,312 delegate votes needed for the Democratic nomination, and in that short span had addressed at least a quarter of a million Americans and had been seen by hundreds of thousands more. The tour had taken its toll on him; he was stricken with laryngitis in the late stages and needed a doctor's treatment, and by the end he was taking twenty pills a day for that and a cold. But he felt he was finding his way. "This trip has convinced me that the American people want to move toward policies of reconciliation here at home," he said in Phoenix. The makings of a new political coalition disenchanted with the status quo had been sampled and found valid: the poor and the minority groups alienated from the affluent; young issue-oriented America turned off by inability to reach their elders and by loss of faith in them; the great middle class alienated by a war that was sapping its newly acquired affluence and its sense of moral justice. The question was how to put the coalition together within the context of a closed political process that for years had automatically given the incumbent President a second term for the asking.

That question soon was answered in a way that rocked the embryo Kennedy campaign to its foundations, sending it soaring with initial euphoria, then pulling it to earth with a realization that the road ahead might be no less difficult.

The Kennedy campaign plane was winging eastward for New York on Sunday night, March 31, when Lyndon B. Johnson took to nationwide television to make a special address. The advance speculation was that he was going to announce yet another troop escalation for Vietnam—not the rumored 206,000, but something less than that number that would be high enough to keep the hawks happy and low enough not to get the doves in a new uproar. But the speculation for once was dead wrong. As the Kennedy plane parked at the terminal at John F. Kennedy Airport, John Burns, the New York Democratic chairman, and Dall Forsythe, a Kennedy aide in New York, rushed up the ramp and inside. To an unbelieving Kennedy, Burns blurted: "The President is not going to run!"

The candidate was just leaving his seat in the right front row of the forward first-class section when Burns' words hit him. He sat down again with the state chairman and with Dutton as Burns, his voice trembling, sought to explain what had happened, what exactly Johnson had said. As Kennedy weighed how he ought to respond, Dutton suggested he make no comment until he could learn more in detail. Kennedy agreed. With his disbelieving wife at his side, he left the plane and pushed his way soberly through a crowd in the terminal, as a woman kept screaming at him, "You're our next President!"

During the drive into Manhattan, through a spring haze that put halos of vapor around the streetlamps, an air of suppressed electricity filled the car. The candidate and Ethel sat silently in the back, with Dutton and Dick Dougherty of the Los Angeles *Times* in the front, occasionally offering observations that hung in the tense atmosphere and failed to break it. Only once did Robert Kennedy make any significant comment. "I wonder," he remarked out of the darkness of the swiftly moving car, "if he'd have done this if I hadn't come in."

CHAPTER FOUR

Indiana

QUEST FOR AN ISSUE

IN THE KENNEDY apartment at 870 United Nations Plaza, Ted Sorensen had been watching the President on television while awaiting his candidate's arrival. The opening moments were gratifying. "We are prepared to move immediately toward peace [in Vietnam] through negotiations," the President had said. "So tonight, in the hope that this action will lead to early talks, I am taking the first step to de-escalate the conflict. We are reducing—substantially reducing—the present level of hostilities, and we are doing so unilaterally and at once."

Sorensen was totally unprepared, however, for the closing remarks: "I have concluded that I should not permit the Presidency to become involved in the partisan divisions that are developing in this political year. With American sons in the fields far away, with America's future under challenge right here at home, with our hopes and the world's hopes for peace in the balance every day, I do not believe I should devote an hour or a day of my time to any personal partisan causes or to any duties other than the awesome duties of this office—the Presidency of your country. Accordingly, I shall not seek, and I will not accept, the nomination of my party for another term as your President. . . ."

Sorensen still was pondering the political impact of these words when Kennedy came in, serious and in a mood to get to work. By this time, a goodly group of insiders was there and more were coming, drawn to the center of decision making by the bombshell. Ethel and Dutton had arrived with the candidate; Mankiewicz had caught a plane up from Washington; Walinsky, Greenfield, Gwirtzman, Schlesinger, Burns and vanden Heuvel joined them. But as usual with Robert Kennedy, there was no mass seminar; instead, the candidate threw out something like, "Well, we can forget about Johnson," and plunged into staccato, almost codelike talk with a few of the key political hands, including brother-in-law Steve, Sorensen and Dutton. It was part of the Kennedy style, with both John and Robert, to dispense with long-winded debate in times of political crisis and to draw around the candidate those aides who were so much on the same wave length with him that much could be left unsaid, or only half-said. In the Kennedy inner circle, the unfinished sentence was commonplace, but completely adequate to convey the full thought. As with many other political tacticians, the conversations often left out names, so that it might seem to the outsider deeply conspiratorial; in the best tradition of Casey Stengel, two insiders could talk on at length, in half-sentences and with references only to "this guy" or "the other guy," and neither participant would ever be in any doubt as to what the other was saying. It was not surprising, in this milieu, that those who tended to the forensic or the philosophical got short shrift. Talky liberals like Chester Bowles were not in the Kennedy style; one of Robert's favorites was Burke Marshall because, as another insider put it, "talking to him was like talking to a sphinx."

Thus, while others in the room discussed what the Johnson shocker would mean to the campaign and how their candidate would have to adjust to capitalize on both the withdrawal and the de-escalation of the war, Kennedy and his closest political aides concentrated on filling the immediate vacuum. The candidate phoned brother Teddy at the Marott Hotel in Indianapolis to check signals, and then the two brothers and brother-in-law Steve, each armed with a list of key Democratic leaders totaling perhaps sixty or seventy among them, started making calls. The focus was on the influential men in the nonprimary states, where already the Kennedys knew that Johnson's pullout would leave the biggest hole—and where Vice President Hubert Humphrey, now a near-certain candidate, would look to party regulars for his greatest strength. One or the other of the three called friendly governors like Harold Hughes in Iowa, Philip Hoff in Vermont, Kenneth Curtis in Maine, neutrals like Warren Hearnes in Missouri and even pro-Johnson men like Richard Hughes in New Jersey. All the powerful mayors, including Daley, and other state leaders like Unruh and some who were not so friendly, were called. All got basically the same conversation: "What do you think? We'll have a chance to talk soon, but for now we just want you to know we're thinking of you, we need you." Direct commitments were not asked, but the seed was planted.

One old political friend whose position was directly affected by Johnson's announcement was high on Robert Kennedy's own list to call—Larry O'Brien, the key tactician of John Kennedy's 1960 campaign and now the Postmaster General in the Johnson Administration. He had been committed to Johnson's reelection campaign but now he obviously would be a free agent.

"I need you badly, Larry," Kennedy told him right off. "We have only four weeks to put Indiana together. We had eight weeks in West Virginia." (Actually it was five weeks until the Indiana primary—the same period that elapsed between the Wisconsin and West Virginia primaries in 1960. But it was true that the planning and organization were much farther advanced then.) Kennedy promised the Postmaster General a free hand in organizational matters and launched immediately into a political nuts-and-bolts discussion of his own strengths and weaknesses in the state, and where the O'Brien mastery could help. He assured O'Brien he would do all he could "to make it comfortable" for the leading Democratic politicians to accept him, by doing a solid political job in the primaries, showing he would strengthen rather than weaken party ties and chances.

O'Brien had not thought it wise for Kennedy to have gone into a state like Indiana. But of course it had not been his place to counsel a man seeking to depose the President in whose Cabinet O'Brien still was serving. Kennedy had approached him earlier, but was aware of O'Brien's difficulty in breaking away so didn't push. Among some of the old 1960 Kennedy hands, O'Brien's decision to stay on with Johnson had not gone down well, but Robert Kennedy did not make an issue of it and he remained on good terms with O'Brien. They never had been extremely close—certainly not the way Larry and John Kennedy had been—but they were two Massachusetts Irishmen who knew their politics and practiced it well. So now, with Johnson out, it was a natural union. O'Brien said he would sort his thoughts out and let Kennedy know.

Before the night was over, though, the Postmaster General had two offers to mull over; another Democrat directly affected by Johnson's decision, Hubert Humphrey, already was projecting the development into his own Presidential candidacy, and he too called Larry. O'Brien numbered Humphrey among his warm friends and held him in high regard as a man and politician. He said he would consider that offer as well.

In the end, the pull to the Irish Mafia and the challenge of another Kennedy campaign won out. Not insignificant in the decision, however, was O'Brien's own disenchantment with the Vietnam War and with the self-delusion about it in the Administration of which he was a key figure. He had labored under two Presidents to build a constructive program of domestic legislation and he could see it slipping away, or at least failing to approach its potential, because of this war. Several days after the Sunday night call, O'Brien notified Kennedy he would come aboard as soon as he could submit his resignation to the President in a courteous and diplomatic way.

In Apartment 14F at the UN Plaza address, the talk had turned to the matter of a public response to Johnson. While Ethel brought out a bottle of scotch and ice, the wordsmiths got to work. Sorensen started drafting the statement on a long yellow legal pad, and soon Walinsky, Schlesinger, Kennedy and others joined in. Sorensen suggested that Kennedy request a meeting with Johnson; it would establish the Senator as a force for unity in the party, and it might help neutralize the gains among orthodox Democratic politicians that Humphrey might expect from Johnson's withdrawal. Kennedy acquiesced.

The telegram, calling Johnson's action "truly magnanimous," finally was dispatched in the early hours, and the meeting broke up, with most of the insiders departing in a buoyant mood. But Kennedy himself, while pleased too that Johnson was out, was well aware that the development brought with it some problems for a candidate who had been campaigning primarily and strenuously against an unpopular President and his pursuit of an unpopular war. From well-placed sources outside the government, the Senator had learned before the Johnson speech that some move toward peace talks was under consideration; but the personal touch at the close came out of the blue. It was a problem of adjustment, but certainly an acceptable one. As with most major policy decisions in these frenetic days, there was little time for any full-dress strategy meeting. Kennedy himself had to grope his way. It was not one of those situations where a hard yes-or-no decision would resolve things. Just two days ahead was the Wisconsin primary; what the temper of the country would be toward Johnson now—and toward the man who had so vigorously sought to bring him down—could only be guessed at. One thing was certain. The time had come to cease fire against him, and to demonstrate goodwill. Anything else would only fan the talk of vendetta and conjure up anew the ruthless image.

The next morning at the Overseas Press Club in midtown Manhattan, Kennedy subjected himself to the political kamikaze of a New York press conference: hot lights, screaming photographers and television quasi-celebrities vying for attention and recognition. The circumstances called for a dignified and cool first exposure after the Johnson withdrawal, and Kennedy did his best to meet that requirement in spite of the heavy obstacles thrown in his path. Calm and unhurried in a dark blue striped suit, Kennedy read the telegram expressing "fervent hope" the new efforts toward peace would succeed, and "respectfully and earnestly" requesting a White House meeting "as soon as possible to discuss how we might work together in the interest of national unity during the coming months."

Ducking questions about the possible peace talks as "inappropriate" at this critical juncture, Kennedy cited the continuing need to bring about racial reconciliation at home. He surprised no one by declaring that "in this spirit, I will continue my campaign for the Presidency of the United States."

To one of the regulars in the press corps who had been with Kennedy from the start, the candidate seemed "drained of the truculence" he had shown toward Johnson, and "dazed by events" that suddenly left him without his prime target. And the President, for his part, was going to do nothing to resurrect that target. Returning from a Chicago speech, he told reporters on the Presidential jet that he would "surely" honor Kennedy's request for a meeting.

That day, Kennedy flew to Philadelphia for a tour of Delaware County suburbs and a stop in Camden. Everywhere the crowds were huge, and it seemed at first blush that with Johnson out of the way, sheer charisma might carry Kennedy to the nomination. In Ridley Township, a car rolled over the foot of a child pressed against it by the crush of the crowd. Barry had to hold onto the candidate so firmly that Kennedy called to him at one point and said: "Don't squeeze me so tight. You'll break my back." At a place called Alpine Inn, a local functionary urged the waiting partisans to give the candidate "a very spontaneous ovation" when he arrived—and they did.

In Camden, without the slightest hesitation, he reverted to his pre-candidacy practice of praising the President. "We take pride in President Johnson," he said, "who brought to final fulfillment the policies of thirty years, and who yesterday sacrificed personal political considerations to win the peace for which all Americans yearn. . . . The peace is above all what they want for the future. They will respect and honor President Johnson, who has sought to take the first step toward peace."

Had the students at Kansas State, and at Vanderbilt, and at San Fernando Valley State College heard these words from the lips of the same man who had roused them with denunciations of Johnson only two weeks earlier, they probably would not have believed their ears. But circumstances were different and Robert Kennedy was a political man adjusting to political realities. Many of the issue-oriented students didn't like that; they looked for truth in a sterilized jar, even in politics—especially in politics; to them, pragmatism was a synonym for opportunism. Armed with that viewpoint, they certainly had the goods on the junior Senator from New York. And without Johnson to shoot at, Kennedy himself became again the target—the young man on the make, not for a cause, but for himself, for the Kennedy restoration alone. It was, to some close aides, his worst and most vulnerable period as a Presidential candidate. McCarthy, typically, had the pithy phrase for it. "Bobby has to shoot straight pool now," he said. "When he was banking his shots off Lyndon it was a different game."

Kennedy, undeniably, was thinking less like the campus crusader he was in 1966 and 1967 and more like the hard-nosed politician he had to be in 1968. He had to subordinate his gut feelings sometimes to the political necessities. As O'Brien put it later: "He became a John Kennedy rather than a Bob Kennedy." It was true, as he demonstrated the next night at a Democratic fund-raiser in machine-politics Philadelphia. Prior to the dinner, Kennedy dispatched O'Donnell to talk with Mayor James Tate, a Humphrey man who only two weeks earlier had called Kennedy "a wise guy" for getting into the race. At the dinner, the candidate referred to Tate as "one of the greatest mayors in the United States"—certainly a novel evaluation—and tried to make light of their differences. "From time to time I'm going to look down at Mayor Tate," Kennedy told the packed convention hall, "to see if he's smiling." Mostly he was not smiling, nor were many of the other ranking hardline Philadelphia pols still tied by political realities of their own or by hawkish views on Vietnam to President Johnson and his obvious heir, Humphrey.

It was receptions like this one—politely cool—that tempered for Kennedy the upbeat mood that the street and the campus crowds projected. In the final analysis the motorcades and the college rallies were Kennedy's crowbar for opening the door to these Democratic power brokers; if they didn't accomplish that, Kennedy was indeed tilting at windmills. Yet, at the same time, as a highly emotional and issue-oriented candidate, he needed the crowds and the reaction himself to keep going. That same afternoon, before the Tate dinner, a mammoth street-corner mob at 15th and Chestnut and an appearance before 10,000 University of Pennsylvania students at the Palestra had energized him. The students were somewhat hostile at first, but he won them with a gutsy, virtuoso performance of "telling them like it was." At those times he was at his very best; jamming down their throats the inconsistency of their opposition to the war and advocacy of student draft deferment while others who couldn't afford to go to college were dying in Vietnam. That attitude really got under his skin; he privately called it selfish and compared it with the attitude of Southwestern fat cats toward the oil depletion allowance. Without going quite that far at the Palestra, Kennedy gave no quarter to the students, and because he was with them on the big issue, the war, he was able to bring them around. His praise for Johnson was not heard here; instead he resumed his criticism within bounds required by the new realities.

"We have to remind ourselves," he told them, "that the war is not over, the bombs are still falling two hundred miles north of the demilitarized zone and reservists are still being called up. Stopping the bombing must be part of a coordinated plan. . . . The first thing we must recognize is that we will have to negotiate with the NLF. It is silly for our government to act as if the NLF does not exist. They are not going to come in just to surrender and give their swords as we did at Appomattox, and be given back their horses and told to go home for the spring plowing."

Late that night, as the Kennedy party flew back to Washington after the Tate dinner, the candidate talked with reporters about the difficult time ahead. The day had provided a graphic example of the Kennedy formula of using public reaction to shake the pols, and of how tough it was going to be. Yet he seemed hopeful that the political ineptness of the Administration, as demonstrated in New Hampshire and to be confirmed that very night in Wisconsin as that state's primary returns rolled in, would help him. "They don't do anything well anymore," Kennedy said at one point.

In Wisconsin, the voters had gone New Hampshire one better. Early returns made it clear that Johnson was going under, without the trace of a sympathy vote to save him. To his Sunday night notice that he was leaving the White House, they seemed to say "so be it," if not "good riddance." McCarthy eventually won 56.2 percent of the 733,000 votes cast in the Democratic primary, compared to only 34.6 percent for Johnson. On write-ins, Kennedy got 6.4 percent and Humphrey 0.5 percent, with the rest scattered among minor figures. And on the Republican side, as expected, it was the unopposed Nixon all the way, his strong organization in the state turning out such an impressive vote that it was becoming increasingly difficult to dismiss him out of hand as a loser.

It was on this rather conclusive political note that Kennedy met with Johnson at the White House the next morning at 10 o'clock. Once again he had his most acceptable second, Sorensen, with him, and this time Johnson had Walt Rostow and Charles S. Murphy, a political aide, in the room. Kennedy and Sorensen sat across from the President at a table in the Cabinet room as Johnson read a teletype message just in. It was the first word from Hanoi, preceded as usual with rampant vitriol, that North Vietnam was willing to start peace talks. He handed the message to Rostow and then to Kennedy, who remarked that it was at least "a first step." For about an hour they talked, discussing the partial bombing halt and its objectives and, eventually, the political situation. It was one of those walking-on-eggs talks, each man deferring to the other, minimizing the differences between them, with the President reiterating what he had said on television—that he would not run, and that the decision was not subject to later reconsideration. Nor would he take sides in the contest for succession within the Democratic Party.

When Kennedy left, Johnson summoned Humphrey and, according to White House press secretary George Christian later, repeated his observations about not running and not playing favorites. "I understand neither of them asked for any specific thing—statements, endorsements, anything of that nature," Christian said, confirming the obvious to reporters who well knew the principals involved. "My impression is that they fully understand the President's position, and there was no talk of a statement or endorsement."

Kennedy returned to McLean for a day of campaign planning. He was headed for a week of intensive effort in Indiana and a string of nonprimary states—Ohio, Louisiana, Kentucky, Alaska and South Dakota—and the pace was so swift that there was little time just to sit down and lay out a master battle chart. Indiana had to be the immediate key; but also essential to the delegate-prying operation in the nonprimary states was the demonstration of public support in every region of the country. So while a skeleton force of organizers was being moved into Indiana, the candidate prepared to take the show on the road. For Indiana, after a scouting trip by Salinger that confirmed the state would be tough, Ted Kennedy's people in Massachusetts were sent in to do the pick-and-shovel organizing. Gerry Doherty, the former Massachusetts Democratic chairman, was to be the day-to-day operations chief, with Kennedy men from Massachusetts, New York and Washington joining him. The national Kennedy headquarters in Washington, now established in an office building at 20th and L Streets Northwest, coordinated the whole campaign, with Steve Smith, Teddy and old John Kennedy hands Sorensen and O'Donnell running the show, plus O'Brien when he came aboard. Key aides identified mainly with Robert Kennedy—particularly the old Justice assistants like Nolan and Douglas—formed another stratum, as did Ted Kennedy lieutenants like Burke, Dun Gifford and Dick Drayne. Citizens groups were organized under another old Justice aide, Louis Oberdorfer, a Washington lawyer, and Walter Sohier, of New York. They recruited from a mail flow of an estimated 5,000 letters a day. The chief speechwriters and researchers like Walinsky, Greenfield, Gwirtzman and Edelman also functioned from here when they were not on the road with the candidate.

For the operational headquarters in Indiana, offices in downtown Indianapolis were leased above an old theater playing *Gone With the Wind*, and the organization started scrambling for volunteers to set up storefronts throughout the state. A fair share of college kids was available to give Kennedy his own version of the now-heralded McCarthy "children's crusade," but finding adult operatives and prestigious figureheads in the smaller farm and rural towns throughout the state proved to be much more difficult. In many places no one came forward, as they had in other states, to accept the honor of being the Kennedy town chairman. Even when the Indianapolis masterminds sent out scouts to dragoon local leadership, they often came up empty-handed. The reason was obvious: Branigin, relieved of the role as Johnson's stand-in for the primary, had decided to stay in the race as a favorite son—in other words, a pro-Humphrey, anti-Kennedy candidate. The move locked up the state Democratic apparatus under St. Angelo and shut off the Kennedy operation from any regular party help. Finally, Ted Kennedy personally, and Larry O'Brien when he joined the team, came into the state armed with lists of Indiana Democrats, holed up in their hotel rooms and began phoning for volunteers. Other Kennedys—Rose, the candidate's mother; Eunice Shriver, his sister; Joan Kennedy, his sister-in-law, and Ethel—toured the state, speaking at teas and trying to drum up grass-roots help. Schlesinger and a fellow writer and native Hoosier, John Bartlow Martin, also pitched in.

The Kennedy's money played its part as well, as Rose Kennedy acknowledged in an interview in which she indiscreetly argued that it was the family's money, so they could do what they wanted with it. Wincing from that blooper, the Kennedy camp went ahead nevertheless and bought television time over most of the state's sixteen outlets. John Frankenheimer, the movie director, and Dick Goodwin when he rejoined Kennedy, worked up television spots and half-hour specials that sought to show the candidate at his mellow best, as a counter to the frantic street campaigning. St. Angelo at one juncture put a $2-million price tag on the Kennedy campaign in Indiana, and the Kennedy camp owned up to $600,000 to $650,000 before primary day. Also, in the course of the Indiana campaign, *Variety*, the show business magazine, quoted Kennedy's advertising agency—Papert, Koenig, Lois—as saying Kennedy would spend $3 million in Indiana, and from $15 million to $18 million in the whole campaign. The agency promptly denied the story, but the impression stuck that the Kennedys were spending freely. It was carefully nurtured by the Indianapolis *Star*, which ran a cartoon picturing Robert and Ethel Kennedy as "Bonnie and Clyde" roaring through the state tossing bills from an open roadster. Whatever the exact Kennedy spending figure, it probably was considerably more than what McCarthy spent. The Minnesotan's finance director near the close of the Indiana primary estimated the McCarthy outlay at about $500,000. Branigin, who had the state organization to work for him, was able to function with much less.

On the morning of April 4, Kennedy returned to Indiana primed for his first full day of campaigning there, ready to start spreading the seed he hoped would flower into his first primary victory a month later. As usual, college rallies provided the kickoff—at Notre Dame and Ball State universities—and the crowds were good if not ecstatic. At Ball State, in answer to a question about racial conflict, he acknowledged there were extremists on both sides of the issue, but added: "Most people in America want to do the decent thing."

In early evening, as the chartered plane was leaving Muncie for Indianapolis, Kennedy for the second time in four days heard some startling news as he sat in the cabin. A reporter who had just called his office rushed up the aisle and told him; Dr. Martin Luther King, Jr., had been shot in Memphis. "Oh, no," Kennedy said, turning his back. The plane was airborne before the reporter could get more details, so Kennedy instructed Dutton to find out two things immediately on arrival at Weir Cook Airport in Indianapolis—King's condition, and the state of the city's Negro wards, in the midst of which the candidate was scheduled to hold a large street-corner rally.

As soon as the plane touched down, Dutton raced to the airport police office, made several calls and dashed back, while Kennedy waited aboard, scribbling some notes. The word was bad—King was dead. As for the Negro wards, they were quiet; apparently the news had not spilled out yet. Ethel was nervous and wanted her husband to cancel the stop and go directly to the hotel. But he said no; he sent her back with an aide, and he and Dutton climbed into the back seat of an awaiting closed car. Barry rode in the lead police car. On the way, Kennedy sat silently in the dark for long minutes. "What should I say?" he murmured, finally. Dutton mentioned only the obvious things about the need to stress nonviolence and faith in racial reconciliation. Kennedy lapsed into silence again.

When they arrived at the rally, it was immediately apparent that the crowd had not heard the news. About 1,000 persons were there, about 70 percent of them Negroes, with plenty of Kennedy banners and the customary gay mood of a political gathering. The lead cars drove as close as they could get to the platform, and Barry came back apprehensively to lead the candidate to it. Kennedy, grim-faced, stepped out into the chilly night, huddled in a black topcoat. He told the local organizer he wanted to speak at once. After a curt introduction, he faced the crowd, eyes moist, and said in a wavering voice:

"I have bad news for you, for all of our fellow citizens, and people who love peace all over the world, and that is that Martin Luther King was shot and killed tonight."

A gasp went through the audience, and some women screamed, "No!" But in the large crowd and the gay mood, not everyone grasped what he was telling them. As he went on, speaking extemporaneously and more movingly than the most veteran of his traveling companions ever had heard him, there were awkward bursts of applause and cheers from the crowd, still not aware of what he had told them. Kennedy himself realized what was happening, and through his own demeanor and the tone of his voice, he eventually subdued them and got the terrible news across.

"Martin Luther King dedicated his life to love and to justice for his fellow human beings, and he died because of that effort," Kennedy said.

"In this difficult day, in this difficult time for the United States, it is perhaps well to ask what kind of a nation we are and what direction we want to move in. For those of you who are black—considering the evidence there evidently is that there were white people who were responsible—you can be filled with bitterness, with hatred, and a desire for revenge. We can move in that direction as a country, in great polarization—black people amongst black, white people amongst white, filled with hatred toward one another.

"Or we can make an effort, as Martin Luther King did, to understand and to comprehend, and to replace that violence, that stain of bloodshed that has spread across our land, with an effort to understand with compassion and love.

"For those of you who are black and are tempted to be filled with hatred and distrust at the injustice of such an act, against all white people, I can only say that I feel in my own heart the same kind of feeling. I had a member of my family killed, but he was killed by a white man. But we have to make an effort in the United States, we have to make an effort to understand, to go beyond these rather difficult times.

"My favorite poet was Aeschylus. He wrote: 'In our sleep, pain which cannot forget falls drop by drop upon the heart until, in our own despair, against our will, comes wisdom through the awful grace of God.'

"What we need in the United States is not division; what we need in the United States is not hatred; what we need in the United States is not violence or lawlessness, but love and wisdom, and compassion toward one another, and a feeling of justice toward those who still suffer within our country, whether they be white or they be black.

"So I shall ask you tonight to return home, to say a prayer for the family of Martin Luther King, that's true, but more importantly to say a prayer for our own country, which all of us love—a prayer for understanding and that compassion of which I spoke.

"We can do well in this country. We will have difficult times. We've had difficult times in the past. We will have difficult times in the future. It is not the end of violence; it is not the end of lawlessness; it is not the end of disorder.

"But the vast majority of white people and the vast majority of black people in this country want to live together, want to improve the quality of our life, and want justice for all human beings who abide in our land.

"Let us dedicate ourselves to what the Greeks wrote so many years ago: to tame the savageness of man and to make gentle the life of this world.

"Let us dedicate ourselves to that, and say a prayer for our country and for our people."

The news, and the eloquent words said with empathy bred of a shared loss, left the audience quiet and weeping. The reference to his brother's assassination was Robert Kennedy's first public mention of it, more than four years later, and of his own reaction to that tragedy. He went straight back to the Marott Hotel, where Ethel was waiting, and phoned Dr. King's wife, Coretta. He ducked into a small bedroom off the living room to make the call. At the close of the conversation, Kennedy asked Mrs. King if there was anything he could do to help her. She said there was; he could arrange to bring Dr. King's body back from Memphis to Atlanta. Kennedy said he would take care of it, and hung up. He had Dutton ask Burke Marshall in New York to fly down to be with Mrs. King, and arrangements were made through Kennedy's office for the chartered plane—a gesture for which he later was criticized on grounds he was seeking political capital from King's death.

The candidate canceled his campaign schedule, then decided to go on to Cleveland to make one more speech the next day. There were calls coming in to him from concerned Negro leaders, some of whom wanted him to do something to help them discourage an outbreak of retaliatory violence. In Washington, D.C., trouble already had started, escalating within the next twenty-four hours into the city's first major riot.

When Kennedy arrived in Cleveland on the fifth, a large crowd was waiting for him in the city's downtown Public Square. But he was late, so he went directly to the City Club of Cleveland, where he talked quietly and again movingly about the crime of Dr. King's murder and about the violence that had overrun the land. It was a speech that could have been delivered, without a word changed, two months later to the day, about an equally senseless act in Los Angeles.

"This is a time of shame and sorrow," Kennedy said. "It is not a day for politics. I have saved this one opportunity to speak briefly to you about this mindless menace of violence in America which again stains our land and every one of our lives.

"It is not the concern of any one race. The victims of the violence are black and white, rich and poor, young and old, famous and unknown. They are, most important of all, human beings whom other human beings loved and needed. No one—no matter where he lives or what he does—can be certain who will suffer from some senseless act of bloodshed. And yet it goes on and on.

"Why? What has violence ever accomplished? What has it ever created? No martyr's cause has ever been stilled by his assassin's bullet. No wrongs have ever been righted by riots and civil disorders. A sniper is only a coward, not a hero, and an uncontrolled, uncontrollable mob is only the voice of madness, not the voice of the people.

"Whenever any American's life is taken by another American unnecessarily—whether it is done in the name of the law or in the defiance of law, by one man or a gang, in cold blood or in passion, in an attack of violence or in response to violence—whenever we tear at the fabric of life which another man has painfully and clumsily woven for himself and his children, the whole nation is degraded.

" 'Among free men,' said Abraham Lincoln, 'there can be no successful appeal from the ballot to the bullet; and those who take such appeal are sure to lose their cause and pay the costs.' Yet we seemingly tolerate a rising level of violence that ignores our common humanity and our claims to civilization alike. We calmly accept newspaper reports of civilian slaughter in far-off lands. We glorify killing on movie and television screens and call it entertainment. We make it easy for men of all shades of sanity to acquire whatever weapons and ammunition they desire.

"Too often we honor swagger and bluster and the wielders of force; too often we excuse those who are willing to build their own lives on the shattered dreams of others. Some Americans who preach nonviolence abroad fail to practice it here at home. Some who accuse others of inciting riots have by their own conduct invited them. Some look for scapegoats, others look for conspiracies, but this much is clear; violence breeds violence, repression brings retaliation, and only a cleaning of our whole society can remove this sickness from our soul.

"For there is another kind of violence, slower but just as deadly, destructive as the shot or the bomb in the night. This is the violence of institutions; indifference and inaction and slow decay. This is the violence that afflicts the poor, that poisons relations between men because their skin has different colors. This is a slow destruction of a child by hunger, and schools without books and homes without heat in the winter.

"This is the breaking of a man's spirit by denying him the chance to stand as a father and as a man among other men. And this too afflicts us all. I have not come here to propose a set of specific remedies nor is there a single set. For a broad and adequate outline we know what must be done. When you teach a man to hate and fear his brother, when you teach that he is a lesser man because of his color or his beliefs or the policies he pursues, when you teach that those who differ from you threaten your freedom or your job or your family, then you also learn to confront others not as fellow citizens but as enemies—to be met not with cooperation but with conquest, to be subjugated and mastered.

"We learn, at the last, to look at our brothers as aliens, men with whom we share a city, but not a community, men bound to us in common dwelling, but not in common effort. We learn to share only a common fear—only a common desire to retreat from each other—only a common impulse to meet disagreement with force. For all this there are no final answers. Yet we know what we must do. It is to achieve true justice among our fellow citizens. The question is not what programs we should seek to enact. The question is whether we can find in our own midst and in our own hearts that leadership of human purpose that will recognize the terrible truths of our existence.

"We must admit the vanity of our false distinctions among men and learn to find our own advancement in the search for the advancement of all. We must admit in ourselves that our own children's future cannot be built on the misfortunes of others. We must recognize that this short life can neither be ennobled or enriched by hatred or revenge. Our lives on this planet are too short and the work to be done too great to let this spirit flourish any longer in our land.

"Of course we cannot vanquish it with a program, nor with a resolution. But we can perhaps remember—even if only for a time—that those who live with us are our brothers, that they share with us the same short movement of life, that they seek—as we do—nothing but the chance to live out their lives in purpose and happiness, winning what satisfaction and fulfillment they can. Surely this bond of common faith, this bond of common goal, can begin to teach us something. Surely we can learn, at least, to look at those around us as fellow men and surely we can begin to work a little harder to bind up the wounds among us and to become in our own hearts brothers and countrymen once again."

The speech was, in a very real sense, a turning point in the Presidential campaign of Robert Kennedy. The views he expressed he had held for years and had been the motivating force of his domestic position. Yet here and now he identified the violence that comes from a government and a people grown apathetic to the needs and the burdens of its poor and its socially shunned— "the violence of institutions: indifference and inaction and slow decay." From this point forward, Kennedy's campaign took on the theme of racial justice and reconciliation to a degree that made Vietnam almost a subordinate issue. With Johnson out and the peace talks soon to start, it is undeniably true that Kennedy, for political reasons, needed a new emphasis. But he was a professional politician who also listened to his heart, and the death of Dr. King, his ally and friend, drove him inexorably in this direction. From here on, the private Robert Kennedy that his closest friends professed him to be—not strident, not bombastic, but low-keyed and sensitive—increasingly emerged in his public life as a campaigner.

Kennedy returned to Washington, walked through the ruins of one of Washington's Negro sections on Sunday, spent Monday at Hickory Hill and then chartered a Lear jet for an evening flight to Atlanta. There he spent most of the night before the funeral in private, unpublicized meetings with Negro leaders, arranged by his old Justice Department lieutenants, Marshall and John Doar. At about 9 o'clock in his hotel suite, he met with about twenty-five men and women less well-known to the white community but influential in major cities around the country, and about ninety minutes later with a group of entertainers gathered for the funeral, mostly Negroes: Bill Cosby, Sammy Davis, Jr., Nancy Wilson, Harry Belafonte, Sidney Poitier. Alan King and Ben Gazzara also were there. Kennedy sat in a corner and listened as they spoke with eloquence, particularly Cosby, about the crisis in race relations. Finally, a third group came in—Dr. Ralph Abernathy, Andrew Young and other leaders of Dr. King's Southern Christian Leadership Conference—and talked on until nearly 3 o'clock. Then, after a few hours' sleep, Kennedy joined the host of other national political dignitaries who attended the funeral and marched in the broiling Georgia sun to Dr. King's burial place at Morehouse College. Like many others who were not famous, Kennedy took off his jacket in the line of march—and was criticized for it by some for not showing proper respect for the deceased. This kind of phony and pretentious public yardstick of grief particularly grated on Kennedy, and baffled him.

The campaign resumed on April 10 in Indiana, on a subdued note. Dick Goodwin by now had left McCarthy and joined the Kennedy effort, on the side of the older pragmatic types who agreed that the candidate would have to tailor his positions to meet the basic conservatism of the Hoosier state. Sharing the same view was O'Brien, who on April 10 announced his resignation from the Johnson Cabinet to join Kennedy in Indiana. There, the candidate struck the first chords of a speechmaking repertoire aimed at the staid small-town vote that in another ten days would have newsmen comparing him to Barry Goldwater in his sudden emphasis on the same economic themes that so gladdened conservative hearts in 1964. He toured Fort Wayne, Columbus and Terre Haute, pressing the theme of curbing "lawlessness and violence" in the streets of the nation, calling for jobs in place of "welfare handouts" and for the return of political power to the local level.

While thus occupied with the conservative rhetoric in Indiana, Kennedy also was required by practical politics to make one last swing through non-primary territory. Starting with Indiana on May 7, there would be six primaries in four weeks—an unprecedented pace for a candidate, and Kennedy could ill afford to lose a single one, or to spend time in nonprimary states once that grind began.

The first stop, Michigan on April 11, provided two newsworthy developments, one of which was the kind of trivia that catches the color of a political campaign and assures front-page coverage throughout the country. The other, the kind that haunts the men who are responsible for the lives of controversial public figures, was known to only two or three Kennedy insiders.

The private development occurred first, at the Jack Tar Hotel in Lansing, situated on a corner directly across from the State Capitol. Kennedy, having met with Michigan Democratic leaders to discuss his prospects for a portion of the state's 96 delegate votes to the national convention, was in his suite getting ready to go to the airport. While he was in the bedroom, Bill Barry came up and told Dutton, who was waiting in the living room, that local police had spotted a man with a rifle on a nearby rooftop. Dutton, not wanting to upset Kennedy, walked casually into the bedroom, went over to the window and drew the curtains. Kennedy, slipping on a clean shirt, looked up at once and said, "Don't close them. If they're going to shoot, they'll shoot."

Dutton said nothing. When it was time to leave, he led the candidate into the hotel elevator, down past the first floor and into the basement, where his car was waiting. "What's the car doing down here?" Kennedy wanted to know. "Well," Dutton told him, "we have a report—maybe serious."

Kennedy frowned. "Don't ever do that," he told his closest lieutenant with some annoyance, climbing into the back seat. "We always get into the car in public. We're not going to start ducking now."

The driver, briefed beforehand to make a fast exit, roared up the exitway onto the street level. "Stop the car," Kennedy told him. He then jumped out and proceeded to shake hands with the crowd outside the hotel for three or four minutes, while Barry and Dutton waited in extreme trepidation. Then the candidate got back in and the party drove off. "He never said a word," Dutton recalled later. "But we got his message." (When Kennedy was not flirting with death, he was joking about it. This same week, after an aborted takeoff in which the campaign plane nearly ran off the runway before stopping, he observed to the accompanying press corps: "I want to say in all modesty that if we don't make it this time, you fellows are going to be in the small print.")

The public development that made news that day occurred in Kalamazoo, where Kennedy was mobbed in a downtown rally. As he finished speaking in the city's Central Square, he bolted for the car, a flock of fans in pursuit, and climbed on top of the convertible. As the car moved out, a twenty-seven-year-old mother of five, Mrs. Phyllis Dean, hoisted herself onto the back next to him and reached for a souvenir. Kennedy felt something tugging at his foot, looked down and saw the woman busily separating him from his footwear. As she told reporters later, "I couldn't get a hanky or anything, so I just took his shoe." It came off still tied, and thanks to Mrs. Dean, the world learned the next day that Robert Kennedy wore a size 8½ with an arch support made in England. (The news overshadowed word from Augusta, Maine, that Governor Kenneth E. Curtis had joined his New England neighbor, Governor Philip H. Hoff of Vermont, in support of Kennedy's candidacy. It was encouraging, but neither of these two young governors could be classified as Democratic power brokers.)

For the next ten days, the Kennedy campaign roamed across the American countryside like some purposeless road show, just picking its way at random. Actually, of course, the design was to demonstrate a nationwide demand for the Kennedy candidacy, and to provide the personal fuel for igniting citizen participation and interest in later primaries. The candidate spent two nights and a day in West Virginia, evoking memories of his late brother's critical primary campaign and victory there in 1960. In the town of Oceana, when the caravan stopped for cold drinks and hamburgers, a resident rushed up with a mugful of home brew, the mug not having been used since a day in 1960 when John Kennedy sampled some of the same concoction. Robert Kennedy drank from it and the valued souvenir was put away again.

After Easter at Hickory Hill, where a photograph with Ethel and all ten children was taken that later was to be cherished in his own family, Kennedy stopped in Indiana long enough to help Polish-American voters celebrate "Dyngus Day"—commemorating the death of St. Stephen, Christianity's first martyr—and to take a parting shot at the Indianapolis newspapers, which were just warming up to their campaign against him. After eating kielbasa and drinking beer at the West Side Democratic and Civic Club in South Bend, Kennedy sang a Polish song he had learned during his 1963 visit to Poland and said of that trip: "The Polish government refused to advise the people that we were there. It's sort of like the Indianapolis papers when I'm here."

In Gary, where the motorcade next took him, the blackout of the Indianapolis papers made no difference. Crowds lined the streets, with kids of all ages racing their bikes alongside Kennedy's car or running as fast as their legs could carry them. Barry, watching this phenomenon, noticed a boy about ten years old, his one arm in a cast, tugging his sister, about four, behind him as he dashed along. Suddenly, when the girl faltered, the boy swept her up onto his shoulders, without breaking stride, and continued after the car. Barry pointed them out to Kennedy. He had the car stopped, lifted them into the convertible and sat one on either side of him. The first thing that occurred to the little boy, named Michael, was to reach over, take his sister's face in both his hands, turn it up toward Kennedy's, and say: "Look, here's Senator Kennedy." This new trio of friends then proceeded for several more miles in the motorcade. Kennedy kept asking Michael where he wanted to be let off; the boy, thoroughly enjoying himself, told his host not to worry, he and his sister would get home somehow later. Finally, Kennedy had his car pulled out of the motorcade. They drove to the children's home, a small frame house, where their mother, in a housedress, served iced tea as Kennedy sat on the front steps and talked, while the motorcade waited.

That night, it was off to North and South Dakota, Nebraska, Oregon, California, Colorado and Nebraska again, showing the flag in the nonprimary states, rallying the troops in those with primaries. The crowds were large, but the results were mixed and unpredictable.

In Coos Bay, Oregon, a local booster group called the Pirates surprised and dismayed the candidate by informing him before a roomful of people that it was their custom to take blood from all prominent guests and to preserve it on parchment. Before Kennedy knew what had hit him, a razor blade had been applied to his finger and blood was dripping from it. "I'm thinking of reassessing my position," he said, in genuine disbelief. "I know you're for Nixon, and you came down to see me have something stuck into me to see if I could survive it."

(This wasn't Kennedy's only trial in Oregon. He was appalled at the slow start toward a real organization in the state for the May 28 primary.)

In San Francisco, a minority in a crowd of 6,000 students in the University of San Francisco gymnasium hooted him, shouted "Fascist pig!" at him as he pushed through the crowd and refused to let him speak. He finally threw away his prepared text and invited questions. One excited young man among the protesters raced down the aisle toward him, causing both Barry and Dutton to move in front of Kennedy. But the young man merely wanted to assail him with a question. "Why," he asked, almost beside himself in excitement, "did you go to work for Joe . . . Joe . . . Joe . . . Kennedy?" "Because," the candidate shot back with a straight face, "he's my father. Next question?" Much of the young crowd apparently missed the questioner's intent to ask about Kennedy's association with Joe McCarthy. But those who didn't, especially those in the press section, roared at the comeback.

In supposedly sedate Scottsbluff, Nebraska, extreme Vietnam War critics at Hiram Scott College heckled him. To an open supporter of the Viet Cong, Kennedy said, "You should do the same thing I did and cut your hair." And when the razzing continued: "Why don't you get bored and go home?"

Some reporters, by this time, began to see signs that Kennedy himself was getting bored, and concerned. The crowds were large as always but the reaction seemed bland. In the Plains States and in affluent Oregon, where there were few Negroes, the preaching of racial reconciliation and of "putting the country back together again" seemed not to fire much passion in citizens who in their placid existence were not aware that it had fallen apart. In fact, all that talk seemed to frighten some folks. After the two-day swing through Oregon, Dick Harwood spotted a danger signal for the Kennedy campaign that some key Oregonians apparently didn't recognize even later. "The difficulty," he wrote in the Washington *Post*, "is that more often than not he is talking to affluent people who have never seen a ghetto, an Indian reservation or a coal camp in Eastern Kentucky. What often happens is what happened Thursday at a suburban shopping center outside Portland. In the middle of Kennedy's speech on unmet social needs and the racial crisis, two reporters noticed something about the crowd. Everybody seemed to be smiling."

Thus it was that Kennedy, with the combatant's zest for a fight, returned to Indiana on April 22 for a near-solid two weeks of campaigning through primary day, May 7. The dimensions and the specifics of the problem were clearly understood by the candidate and his chief strategists: a conservative state, hawkish on the war, with a 9 percent Negro population; a significant blue-collar white vote in industrial Lake County, which had voted the back-lash for George Wallace in 1964; an influential press committed hard against him; a strong state Democratic Party working to defeat him and denying him usual precinct organization; a three-man race in which the antiwar constitu-ency might be split with McCarthy. All in all, it was quite an order for a fresh-man Senator from the East, even a Kennedy—especially a Kennedy in Indiana.

On top of all that, there now was a challenge from McCarthy, made in a radio interview in Pittsburgh five days earlier, to debate the issues on television. The idea did not fit in with the Kennedy plan: to ignore McCarthy, let him fade into obscurity, dispose of him early, pick up his antiwar support and move on to a showdown against Humphrey in the nonprimary states before the August Democratic convention. But McCarthy said he already had received an invita-tion from one television outlet and had indicated he was willing. "I think this would be a very good test between Senator Kennedy and me," McCarthy said. In the spotlight before Kennedy's entry into the race, McCarthy was being lost in the blur of the Kennedy campaign of charisma, and he needed the exposure and the direct shot at his new foe. "I think this kind of joint appearance is more helpful in the primaries," he said, "where contrast of persons is more impor-tant than in national elections where there are party positions and issues." But Kennedy ignored the debate challenge. He was concerned with how Indiana Democrats saw him, not about how they compared him with McCarthy.

Kennedy already had moved to identify his position on the Vietnam War as more moderate than it had been painted; everywhere he had stressed right off he did not favor unilateral withdrawal (neither did McCarthy). But with that issue receding as a result of the peace talks, and the matter of racial unrest— "crime in the streets" was the prevailing code phrase—emerging as the critical problem, Kennedy was sitting squarely on top of a political powder keg. He was closely identified with the cause of Negroes, overwhelmingly so, and that was a strength. But if that identification drove off the "ethnics" in Lake County, that advantage would be canceled out. And if his chances with rural conservative whites likewise were undermined by his association with the blacks, he could pack up and go home.

Required here was the cementing of a difficult but potentially powerful coalition: the blacks with the whites who most feared them, the blue-collar ethnics who shared industrial jobs and communities with them. They had their differences and their deep antagonisms and tensions, but they also had a common lot; in many cases they shared poverty, and if not poverty, then a disenchantment with not having as big a slice of affluence as other Americans. Reduced to the shorthand of politics, they were the have-nots, and if they could be pulled together in an economic coalition against the haves, Kennedy would have his necessary political base.

At the same time, if rural and small-town conservatives could be made to appreciate that the alternative to racial violence was the disarming of the have-nots through broader job opportunity, that significant segment of voters could be hooked into the have-not coalition. This was especially so if the traditional conservative virtues of private enterprise and local government could be plugged in, too. Thus, for example, in deploring "welfare handouts" and calling on private business and local government to substitute jobs and job training for the disdained "dole," Kennedy could make points with everybody. That was his objective; that was what politics was all about. And, happily for Kennedy, it all coincided with his deep belief that progress could come only through racial reconciliation, and with his moral and emotional commitment to that end. It was good politics, and it was how he felt; he embraced the strategy with both confidence and conviction.

Some of the Young Turks in the Kennedy organization, such as speechwriters Walinsky and Greenfield and researcher Edelman, were disturbed about the rhetoric of old-line conservatism creeping into their candidate's campaign pitch. They questioned the need and the wisdom of talking about "law and order" and the inevitable implications of repression the phrase evoked; to them, the clear answer was greater understanding among whites of the plight of the black man and his provocation to violence, and remedial action. Walinsky particularly agitated against the conservative rhetoric. As a bridge to new ideas and to the new politics, he was extremely valuable to Kennedy, and extremely influential in shaping his thinking about Vietnam and social injustice at home. But Kennedy did not count him among his political insiders, and for the good reason that he was inexperienced in the rough-and-tumble of practical politics. He was not tuned in on the early meetings to consider a 1968 candidacy precisely because it was being weighed as a political, not an issue, decision; so it was in Indiana and the later primaries, where getting elected was an essential—the essential—element. Negroes, Kennedy emphasized, were hurt more than whites by crime, so he had to talk about it. "We're going to talk about what people will listen to," Kennedy said of Indiana. The rural whites, he said, "don't want to listen to what the blacks want and need. You have to get them listening by talking about what they're interested in, before you can start trying to persuade them about other matters."

This philosophy guided Kennedy as he plunged into the critical last two weeks in Indiana. On the plane from Washington, he told two newsmen that the next three days would take him to the toughest areas he faced—the conservative strongholds of the state, populated by farm Republicans and Democrats up from Kentucky and Tennessee. "These people don't like Negroes; they 'like' the war; they don't like the kids," Kennedy said. "Because of my Negro support, I have a problem McCarthy doesn't have. And Branigin is a problem because I can't attack him; he's a Hoosier, and he's theirs. My advantage is he's not a serious candidate, but that's all I can afford to say about him."

With all these headaches, why go into Indiana at all, Kennedy was asked. "I had to have a fight somewhere, and soon," he answered. "Nebraska didn't look too good and that was only a week later. I needed an impressive thrust right off. Branigin was supposed to be just a stand-in when I filed, and St. Angelo told me he tried to get Branigin to stay out. But now I'm in. It will be tough. In Kokomo, we tried six places to set up a headquarters, and they all turned us down when they heard it was for me."

In gauging the effectiveness of a Kennedy campaign, it always was necessary to guard against "poor-mouthing"—the art of telling it worse than like it is. O'Brien, Dutton, Salinger, Sorensen, Kennedy himself, all were past masters, and the early strategy in Indiana was to paint Kennedy as an underdog who would be lucky to finish second. But the next three days of campaigning indicated that Kennedy's evaluation of the task was not overly exaggerated. Although crowds were good, curiosity seemed to be as much the motivating factor in turning them out as enthusiasm for his candidacy. The first day was devoted largely to a respectful tour of Hoosier shrines, to enable the Eastern candidate to demonstrate that he appreciated the state's role in building the country. Three of the Kennedy children were along—David, twelve; Courtney, eleven; and Michael, ten—as well as Ethel and Freckles, the family springer spaniel. The Senator lost no opportunity to tell the Hoosiers who turned out for him how impressed he was with the historical landmarks and what an education it was for his three kids.

Actually, it was a drab day. Freckles probably caused the most excitement for the Kennedys when he wandered into the Old Cathedral in Vincennes, the first stop. The venerated church, supposedly the oldest in the Old Northwest Territory, struck no awe in Freckles, who never was subjected to leash, even in the wildest crowds. Dutton, standing near the back of the church while the Kennedys examined some of its features, looked up at the altar and spied Freckles casually strolling around on it, sniffing. "Oh, my God!" Dutton exclaimed, in a tone of horror that could not be mistaken for a churchgoer's routine supplication, as he raced up to fetch the dog.

Kennedy had an early opportunity to test his conservative rhetoric at a luncheon sponsored by the Civitan Club of Vincennes. The members and guests were restive and in a skeptical mood about this young Eastern whippersnapper who had blown into their state to make political capital of it. But what they heard must have surprised many of them. For line after line, he fed them the straight conservative gospel, to show he was not the wild-eyed Eastern liberal and free-spender they may have thought he was.

"I was part of an Administration in which private enterprise and the people were freed from the cycles of boom and bust that had plagued this country for six generations," Kennedy reported. "The profits of corporations after taxes rose during those years by almost forty percent, and small business shared in this prosperity. . . ."

"Today, business can extend its work to the unfinished business of our country. There is no commandment which says that government must undertake these tasks. . . . That is why I have introduced legislation to lower taxes to private enterprise, which will undertake programs to wipe out hard-core unemployment and provide housing. That is why I believe the most important step we could take in ending poverty in America in our towns and farms, as well as in our cities, is to involve the private enterprise system. . . ."

He went on like that, laying it on thick: "It is the federal government's job to help; to provide resources; to assure the money is honestly spent; and to demand performance. But it is for the people themselves to control the programs that will change their own lives—whether in the towns of Indiana, the Delta of Mississippi or the ghettos of New York. . . .

"The primary task of the next Administration will have to be the formulation of the fiscal, monetary and foreign policies that can bring the cost of living under control. This must include cutting out unnecessary government expenditures and the reforming of our tax system, so that we can cut the budget deficits. . . .

"It is not likely that speculators would be threatening the stability of our dollar—if we could end a struggle abroad that costs us thirty billion dollars every year. It is not likely that government would find itself with ever-higher budgets and social progress slackening, if it understands that private resources and community control are a far more efficient use of resources rather than simply enacting more and more centralized programs which do not do the job. It is not likely that businessmen would suspect all government effort—if the government enlisted business as an ally of social progress."

With that kind of Wall Street litany, one might have expected the audience to erupt. Instead, it remained polite—and unconvinced. In the question-and-answer period, Kennedy fared no better, even when a question on law and order produced the reply that "I was the chief law enforcement officer of the United States. . . . The law has been my life . . . I don't believe in violence." And then there was a question about the anti-rat legislation in Congress. For the moment, Kennedy put his conservative battle plan behind him: "Do you know there are more rats in New York than people, and there are nine million people there? Children spend their nights trying to keep rats from biting them. . . . You can smile. It's true . . . we're not going to tolerate the riots, we're not going to tolerate the lawlessness, we're not going to tolerate the violence. But we're going to do something about the conditions." Reaction: zero. So it was back to the litany: "I think it's a mistake for the federal government to decide where a school should be located, where a road should be located. . . . We have to

strengthen our police departments so they know how to cope with riots. . . ."
Again, no response. Kennedy left the Civitan meeting wondering what it would
take, after feeding them their favorite economics clichés, to get to such people.
"I know most of you are Republicans," he began at one point, and then, apparently thinking better of it, broke off.

In the town of Washington, the next stop, it was more of the same. A large
crowd turned out for the rally at a main intersection, and kids jumped and
screamed—while their parents stood on the fringe and just watched. In dealing
with local problems, he told them, the federal government "must not take over,
but when there is an acute need, the federal government can come in and help."
Polite applause. Mrs. Jessie Swann, the widow of a local farmer, watched the
performance and remarked, "He's got oomph, all right. But we're Republicans
here. We loved his brother, but this is different. We're for Nixon, but the Kennedy family has a lot of glamour. I wouldn't have missed coming for anything."
Others were less charitable. David Broder, writing in the Washington *Post*, observed that "in almost every crowd . . . there are the Kennedy-haters—men and
women who tell explicitly and uninhibitedly what they would like to see happen
to that blankety-blank." Still others admitted to being vaguely disturbed by the
excitement Kennedy stirred among the kids. It was partly to overcome this impression that he went into small towns. But everywhere it was the same. Once,
driving to a small town and discussing this very problem, a school door opened
and hundreds of kids poured out, stopping the car. As Kennedy got out to talk
to them, he handed Dutton a note. It said: "I hate children."

That was the way it was all that day in Indiana—good crowds by ordinary
standards, if not by Kennedy standards—but uncertainty about their meaning. In the earlier screaming, pushing mob scenes of the Kansas campuses and
California, it had become a standing gag on the press bus, as the pandemonium unfolded, to ask: "Yeah, but will they vote for him?" Here in conservative
Indiana, it was a valid question. The kids always were with him, but even his
private-enterprise pitch didn't seem to turn the voters on. It was not that the
conservative line sounded awkward coming from his lips, either. Actually, he
had been saying the same basic things everywhere from the very beginning of
the campaign; what was different was the emphasis, the tailoring of the rhetoric of participatory democracy at the local level to old-fashioned do-it-yourself

conservatism. Basically, there really wasn't a great deal of difference between the two. It was true, as he often claimed, that he said the same things in Alabama that he said in Watts, in the Eastern cities and in the Midwest; *how* he said them was what was different.

Kennedy concluded his day in southwestern Indiana with what was to be a big rally in a 12,000-seat field house in Evansville, the Democratic oasis that had enabled John Kennedy to carry the Congressional district in 1960 and Lyndon Johnson to do the same in 1964. But poor advancing and scheduling—or public apathy—produced more empty seats than filled. It was not a good sign, and Mankiewicz took much ribbing from the press corps about it. The next morning on the press bus, en route to Elwood, he took the intercom in hand and announced: "We are approaching the home of Wendell Willkie, the Republican candidate for President in 1940. It has been reported he had a crowd of 100,000 when he started his campaign here. Actually, it was badly advanced, because his speech was given in a stadium that held 300,000."

Advancing on a Kennedy trip seldom was poor, especially if Jerry Bruno was in charge. A veteran of the 1960 campaign of John Kennedy, Bruno's cardinal rule was to select a site for speech or rally that for certain would be filled or—preferably—overfilled. A large stadium or auditorium usually was rejected unless a huge crowd was assured. If the turnout was unpredictable, Bruno would look for a small hall, or a downtown intersection where spectators would have to crowd into small standing space, thus giving the semblance of a throng. The "advance man" was just what the name implied. At least several days before the candidate was to arrive, the advance man would go into the city, contact those public officials or party dependables who would help arrange for motorcade permits, police escort, cars, buses, campaign signs, special telephones for staff and press, and all the other necessary paraphernalia for a successful visit. The best routes in and out of the city would be selected and volunteers organized to turn out the faithful and interest the uncommitted. The Evansville trip had been left to a local supporter, with results that would have reduced the hard-bitten Bruno to tears. But in the Kennedy operation, the same mistake seldom was made twice. For the rest of the Indiana primary, the trips were professionally advanced and lack of crowds hardly ever was a Kennedy problem after that.

The day that started in Elwood was marked by still more conservative rhetoric, interlaced with appeals to Hoosiers not to "waste your vote"—a thinly veiled reference to Branigin. Indiana had a special opportunity to decide who would be the Democratic nominee, and "probably the next President of the United States." Only a few other states had that opportunity, Kennedy said; Indiana could be what West Virginia was in 1960.

By now it certainly was clear that Kennedy hoped to make Indiana *his* West Virginia—the place where his campaign would be ignited, where the active competition would be shaken off, as West Virginia served for his brother eight years earlier. But in West Virginia, John Kennedy found two elements that were not prevalent in Indiana: deep, pervading poverty, and a kind of collective inferiority complex. West Virginia's poverty gave John Kennedy a peg for his entire campaign—and at the same time shook him emotionally in a way he seldom had been reached before. The sense of inferiority as a state provided an opportunity to turn around the religious prejudice that undeniably existed toward the Catholic candidate in many parts of the state. The candidacy of John Kennedy, the word went out, gave West Virginians a chance to prove that they weren't bigots. Whether they were or not, that's the way a defeat for Kennedy in the state would be read outside West Virginia. It wasn't put all that crassly, of course, but the message was there, and it worked.

In Indiana, there just wasn't that kind of poverty, or that sense of inferiority. Except for some of the Negro wards, Indiana citizens were doing quite well, and the religious issue just didn't seem to matter much anymore. And far from there being any collective inferiority complex, Hoosiers seemed to bestow upon themselves virtues and achievements that escaped the eye of the outsider. The sense of state pride, in fact, was downright chauvinistic, and that was the trouble. Also, Robert Kennedy already was emotionally committed to the plight of the poor, having seen so much of it elsewhere. Developing a relationship with the people of Indiana by demonstrating his compassion just was not in the order of things. And beyond all that, as Kennedy often told the old 1960 hands, "1968 is not 1960." If Indiana were to be Robert Kennedy's West Virginia, the case would have to be made purely on the basis of the people against the power brokers. And that wouldn't be easy, because both the power brokers and the people were Hoosiers.

Yet, all was not lost. There was the industrial base—the Negro and ethnic "have-nots"—to be glued together, and there was the fact that Kennedy himself was emerging as a prime issue. The Pulliam newspapers were cranking up one of the most blatant efforts to undo a Presidential primary candidate yet seen, and if Kennedy could demonstrate that he wasn't the ogre he was being painted, he might be able to turn that attack and the Branigin-party organization assault to his advantage. His strong suit was personal campaigning, so there was nothing else to do but get on with it.

To inject a little variety into the campaign, and to give the television crews and the photographers something different, Kennedy decided to do some old-fashioned whistle-stopping through north-central Indiana that same day, April 23, along the route of the old Wabash Cannonball. The candidate and full party got aboard the train, a five-car diesel, at Logansport and churned through the pleasant spring afternoon, stopping for depot speeches at Peru, Wabash, Huntington and Fort Wayne. At Logansport, the local mayor gave Kennedy a send-off while a police officer stood on the roof of the National Bank of Logansport, rifle in hand, and studied the crowd, adjacent buildings and other rooftops.

It was a happy trip, with the candidate in a playful mood. At each stop, a banjo group would belt out "The Wabash Cannonball," the song made popular by the baseball pitcher Dizzy Dean, when he announced games on the radio in the 1950's. Then Kennedy would introduce Ethel and the three kids, tell the Hoosiers who jammed around the observation platform of the last car how much the Kennedy family was enjoying Indiana and the Wabash Cannonball, and urge the voters not to waste their ballots on Branigin. McCarthy, all through Indiana, was the forgotten man as far as Kennedy was concerned.

In Peru, pronounced "Pay-roo" by the locals, Kennedy made much of his opportunity to visit the town, saying in jest, as he did in many Indiana towns, that he had decided to run for President just so he could go there and show it to his children. When some local kids who had climbed the back railing to shake his hand wouldn't let go as the train pulled out, Kennedy exclaimed: "They told us Peru was an exciting city, but I never thought it would be like this!" In Wabash, which calls itself the light bulb capital, a sign showed the home product and proclaimed unabashedly, SOCKET TO 'EM, BOBBY. (The pun was nearly as outrageous as a campus version that took note of Kennedy's penchant for quoting the Greeks. That one said, SOPHOCLES IT TO 'EM.)

At each stop, Kennedy would work his way into his serious message of a new and better America, and would close with a favorite quotation he had been using from the start of the campaign. It was a loose paraphrase from Shaw's *Back to Methuselah,* that John Kennedy had invoked in addressing the Irish Parliament in the summer of 1963. In the Robert Kennedy version, it went: "Some men see things as they are and say, 'Why?' I dream things that never were and say, 'Why not?' " Whenever newsmen heard Kennedy tell his audience, "As George Bernard Shaw once said—" they knew he was winding up and it was time for them to get back aboard. But at one of the train stops, he omitted the cue ending and several members of the press corps were stranded. En route to the next stop, Warren Rogers of *Look* magazine was dispatched as an emissary to make sure the omission didn't happen again. The candidate laughed and agreed: from then on, there were few campaign speeches anywhere that didn't close with "As George Bernard Shaw once said"—and with a mad scurrying for the train, or the press bus, or the airplane. Later on, one observer swore he saw Freckles dash for the lead car of a motorcade at the sound of those words.

While the Senator and Ethel relaxed in the last car and the Kennedy children roamed through the train munching potato chips and drinking soda, several of the reporters who regularly covered Kennedy were busy in the club car composing their own version of "The Wabash Cannonball." Among the lyricists were Rogers, Dave Broder, Dave Halberstam of *Harper's* magazine, Jack Germond of the Gannett Newspapers, and Dave Breasted of the New York *Daily News,* who just happened to have his guitar along. Before the train had reached Fort Wayne, this singularly motley crew had seven verses whipped up, all dedicated to the candidate. He and Ethel came into the press car, a pack of aides behind them, and were serenaded by Breasted as he strummed his guitar and twanged out the ersatz epic:

Oh listen to the speeches that baffle, beef and bore
As he waffles through the woodlands, and slides along the shore.
He's the politician who's touched by one and all.
He's the demon driver of The Ruthless Cannonball.

He came down to Logansport one sunny April day.
As he pulled on through the depot you could hear those Hoosiers say,
He's the heir apparent, full of chutzpah, full of gall.
I'll bet he wants our helpin' hand on The Ruthless Cannonball.

His Eastern states are dandy, so all the people say,
From Boston to Virginny, and New York by the way.
The Blacks in Gary love him, the Poles will fill his hall
There are no ethnic problems on The Ruthless Cannonball.

There goes Roger Branigin, the Hoosier's favorite son.
He doesn't want the office, he only wants to run.
His highballin' days are over, he's riding for a fall.
They're noted for long memories on The Ruthless Cannonball.

Now good clean Gene McCarthy came down the other track.
A thousand Radcliffe dropouts all massed for the attack,
But Bobby's bought the right-of-way from here back to St. Paul,
'Cause money is no object on The Ruthless Cannonball.

Old Hubert's got Big Business, Big Labor and Big Mouth,
Aboard the Maddox Special, a'comin' from the South.
Lyndon's got him preachin', so ecu-meni-call,
But soon he'll be a'heavin' coal on The Ruthless Cannonball.

So here's to Ruthless Robert, may his name forever stand,
To be feared and genuflected at by pols across the land.
Old Ho Chi Minh is cheering, and though it may appall,
He's whizzing to the White House on The Ruthless Cannonball.

The Kennedys took the saga in good spirits, applauding at the end. And when cries of "Speech! Speech!" went up from the reporters, the candidate paused for quiet, fixed his much publicized "cold blue eyes" on the authors, and intoned: "As George Bernard Shaw once said—The same to you, sideways."

Then he walked over and asked them for a copy. One of the lyricists handed it to him and cracked, "Forget where you got it." Kennedy fixed "the look" on him again and said: "Oh, no, I won't." He took a few steps away, turned and grinned. "See, it keeps slipping out all the time," said Ruthless Robert.

"The Ruthless Cannonball" became the press' theme song for the Kennedy campaign, leaving still in the fertile and idle minds of the authors another lyrical embryo that never needed to be hatched thereafter. It started: "I don't want a wabbit in the White House—" a reference, of course, to the Senator's toothy countenance. Sometimes newsmen would compare him—behind his back—to a famous cartoon character. What few of them knew was that Kennedy was well aware of it; as he peered out of the campaign plane at the jumping, screaming crowds waiting for him, he would straighten his jacket, brush his forelock back with his hand, turn to his wife and say: "Let's go, Ethel, here comes Bugs Bunny!"

While Kennedy thus was gamboling through the Indiana countryside, the editorial offices of the Pulliam papers, the Indianapolis *Star* and the *News,* were brimming with barbs against him of a more serious nature. The senior Pulliam was a friend and supporter of Branigin, and the papers were conservative—and Hoosier—to the core. Up to this point, they had been giving Kennedy short shrift in the news columns. But starting on the day after the Wabash Cannonball made its historic run, on April 24, a concerted program was begun against him. For the next six days, the *Star* ran stories very favorable to Branigin on page one, with references to Kennedy and McCarthy, if there were any, tucked well inside. The page-one stories trumpeted labor support for the Governor, his attacks on Kennedy and McCarthy for criticizing President Johnson, his shock at their rate of campaign spending and—though no other newspaper of note found any trace of it—a boomlet for Branigin to be the Vice Presidential candidate on the 1968 Democratic ticket. On three of the days, pro-Branigin cartoons ran on page one, one of which seemed to characterize Kennedy and McCarthy as lechers, or worse. Entitled "Guests in the House—!" it showed Branigin sitting benignly at the dinner table while "Mrs. Indiana" warded off the advances of McCarthy and especially Kennedy, who—if the naked eye did not deceive—seemed to have his hand on her left breast.

To the reader who could not get around the state himself, the impression from the *Star*'s stories was that Branigin was a veritable whirling dervish, cutting a captivating figure through multitudes of adoring Hoosiers. Out-of-state reporters who accompanied the Governor, however, found more times than not that he was a solitary figure who roamed county courthouses looking for hands to shake, talking to assembled party faithful and citizens he met on the street. Meanwhile, McCarthy was drawing fair crowds and Kennedy by far the largest of the three—and winding up in the back pages of the newspaper, or not at all.

Purely in terms of controversial news, Kennedy was the frontrunner, and the papers would have had plenty of bona fide ammunition with which to hit him had they covered his campaign the way scores of out-of-state newspapers did. On April 26, for example, he engaged in a lively debate with medical students at the Indiana University Medical Center, a debate which the *Star* covered—but much later, and in a somewhat circuitous fashion.

At the Medical Center, a group of students in white were challenging Kennedy's views on medical care for the poor. One student questioned the need to increase social security benefits; another said the problems of higher medical costs could be handled by private insurance companies; another suggested ghetto dwellers weren't making use of the medical facilities already available to them. Finally somebody asked: "Where are you going to get all the money for these federally subsidized programs you're talking about?" Kennedy shot back: "From you." And then, leaning toward the students, he began:

"Let me say something about the tone of these questions. I look around this room and I don't see many black faces who will become doctors. You can talk about where the money will come from. . . . Part of civilized society is to let people go to medical school who come from ghettos. You don't see many people coming out of the ghettos or off the Indian reservations to medical school. You are the privileged ones here. It's easy to sit back and say it's the fault of the federal government, but it's our responsibility too. It's our society, not just our government, that spends twice as much on pets as on the poverty program. It's the poor who carry the major burden of the struggle in Vietnam. You sit here as white medical students, while black people carry the burden of the fighting in Vietnam."

Somebody interrupted by shouting: "We'll be going soon!"

"Yes," Kennedy went on, "but you're here now and they're over there. The war might be settled by the time you go."

This remark brought hisses and boos from the audience. Again Kennedy repeated that there were "not many black faces" in front of him.

"Hey!" a Negro student called out.

"I can see you," Kennedy replied, "but you sure stand out."

Then the tension broke. Another student asked Kennedy if he would end medical school draft deferments. "The way things are going here today," he replied, "I'd say yes." The reaction helped him understand, Kennedy said, why his agents were having trouble "forming a doctors' committee for me in Indiana."

Before Kennedy left the hall, a student rose and told him that "a lot of us agree with what you are saying." There were cheers and applause from the crowd of about 800. It was the kind of give-and-take that he most enjoyed, and at which he was most effective. It also was an example of his inability to keep his emotional feelings in check all the time, even when the strategy called for him to woo the conservative vote.

All this would have been fodder for a good story in the Indianapolis papers, but the medical school appearance did not appear until several days later. Then, the *Star* ran a front-page story about charges by a former Democratic national committeeman and Branigin friend, Frank M. McHale, that Kennedy was guilty of "injecting racism and religion into Hoosier politics." McHale was further identified—to establish his expertise on both racism and religion, no doubt—as "a former American Legion national committeeman who was named a Knight of St. Gregory by Pope John XXIII." The racism, McHale said, was Kennedy's statement to the medical students that Negroes carried "the major burden" of the war in Vietnam. The injection of religion referred to a statement Ethel had made before a Catholic women's group thanking the Archbishop for "inviting" her when in fact he hadn't. It was, McHale charged, "a deliberate attempt to make Hoosier Catholics think that Bobby had official sanction." The McHale story, incidentally, was the first Kennedy story of any kind to make page one of the *Star* in six days. That kind of breakthrough, obviously, he could have done without.

While the *Star* labored to find new ammunition to use against Kennedy, he was off on another whistle-stop tour, this time from Wyoming eastward through Nebraska to cover remote areas in advance of that primary. At one juncture, the route dipped briefly into Colorado and through a town few ever had heard of, called Julesburg. Though no speech was scheduled there, the train stopped, and startled onlookers saw Robert F. Kennedy, his wife, aides and newsmen jump off and scamper wildly to the rear of the train, brandishing campaign signs.

The placards said, to the utter befuddlement of the locals. FRED DUTTON'S BROTHER FOR ATTORNEY GENERAL; BEAUTIFY AMERICA, GET A HAIRCUT, FRED; and—with Kennedy holding it high—SOCK IT TO 'EM, FREDDY. Everybody was wearing a "Dutton Button" and following Kennedy's lead as he chanted, "We want Fred!" In a moment, out onto the observation platform there appeared a grinning, surprised Dutton. Julesburg, it turned out, was his birthplace. Ethel and girls on the Kennedy staff had improvised the signs and now were whooping it up for the new candidate.

Dutton gave a short speech, using all the Kennedy promises to "do better" and to "turn this country around," and with cheers in his ears, started to re-enter the car.

"Hey," yelled Ethel from under a bright orange hat with a Dutton sign on it. "How about George Bernard Shaw?" Whereupon Dutton returned and delivered the standard Kennedy peroration. Ethel was pleased. "Let's hear it for George Bernard Shaw," she ordered her campaign committee. And as the cheers settled over quaint little Julesburg, Colorado, the Dutton Cannonball chugged off toward Nebraska.

Amid the gaiety, nobody seemed much concerned that on the same day, in Washington, D.C., still another candidate—Hubert Horatio Humphrey—had declared his availability for the Presidential nomination of the Democratic Party. Humphrey's entry into the race, of course, was taken for granted by the Kennedy forces. Activity in the Vice President's behalf had been under way almost from the moment of Johnson's announcement that he would not run again. Behind the clamor of open campaigning in the primary states, operations just as intense were going on in the political backrooms of the Democratic Party across the country. Ted Kennedy and Ken O'Donnell were taking soundings and making exhortations in behalf of their candidate to key local and state party leaders in the large and small nonprimary states. For Humphrey,

Senators Walter Mondale and Fred Harris, Secretary of Agriculture Orville Freeman, Mayor Jim Tate of Philadelphia, long-time Humphrey lieutenant Bill Connell and aides of labor chieftain George Meany all were working overtime on the phone and in person to tap the same convention-delegate lodes. Governor John Connally of Texas was leading a favorite-son strategy among the governors to keep the Kennedy delegate-hunters off the reservation in their states, and Dick Daley remained uncommitted against the probings of both sides. It was clear from the outset that the nonprimary states would be another tough hurdle for Kennedy, even if he should make his way unscathed through the six primaries facing him.

Returning to Indiana, Kennedy found that the Indianapolis *Star* somewhat miraculously had found the ideal running mate for Humphrey. He turned out to be, of all people, Governor Roger D. Branigin of the State of Indiana. Branigin was, of course, all the things a Vice President could not be: too old (sixty-five); too free-swinging in a campaign for the Presidential candidate to risk; from a state too small to deliver a critical bloc of electoral votes and too favorably disposed to Richard Nixon to go Democratic, anyway. But none of these obvious points prevented the *Star* from reporting that, according to "a spokesman in the Humphrey high command," Branigin was "now being frequently mentioned as a possible running mate, along with three or four others." That afternoon, the *News* also spread the word: BRANIGIN A TOP CHOICE FOR VICE PRESIDENT.

Kennedy, undeterred by the revelation of Branigin's new national prominence, began his final full week of Indiana campaigning. The polls now showed him nearly twenty points ahead of the Governor and in even better shape against McCarthy, and so he stuck close to his conservative line. He visited the city of Richmond, only recently the victim of a downtown gas explosion, and cited that community's cooperative rebuilding effort as the kind needed to cope with nationwide domestic ills. Not only did he reject the old New Deal of the forties, he expressed doubts that the policies of the sixties, presumably including those of his late brother, would do anymore. He called for a "new philosophy of the Democratic Party" grounded in bringing government back to the people at the local level.

And in a rousing close to a question-and-answer period at Valparaiso University Kennedy turned around a student inquiry on federal assistance: "How much time do all of you spend, over the summer or on vacation, working in a ghetto, in eastern Kentucky, or on Indian reservations?" he wanted to know. Instead of asking what the federal government is doing about starving children, Kennedy said, "I say, why don't we do something about it?" He called on the students to "organize yourselves here" to help the starving in America, and he closed with his favorite Camus quotation about relieving the plight of tortured children. The applause was deafening.

That afternoon Kennedy went on to the Gary-Hammond area where a motorcade took him through city streets jammed with Negroes, and on to St. Michael's Ukrainian Hall. These were, in the words from "The Ruthless Cannonball," the blacks who loved him and the "ethnics" who filled his hall, and they demonstrated the truth of the lyric in an impressive outpouring. It was a surprising coalition in supposedly backlash country, this black-blue collar coalition of the have-nots, but it seemed to be in the making.

From the Ukrainian Hall, Kennedy made one more stop, a rally in East Chicago sponsored by a Spanish-American "Viva Kennedy" Club. As usual, the motorcade, slowed by the crowds, arrived more than an hour late, and the waiting Latin-American audience had become impatient and unruly. There were great cheers for Kennedy as he moved to the microphone on the floor of a large high school gymnasium, but the chattering in the stands was incessant and the moving about constant and distracting. It had been a tremendously long, tiring and effective day; all this crowd really had wanted was to see Kennedy, yell a little and go home. He obviously had plenty of votes here. But he wanted to reach this crowd, have them react to him, and the more unruly they became and the more difficult to corral, the harder he worked at it. First he tried joking with them, then orating at them, then, finally, playing games with them—speaking their native language in a halting, childish, eventually condescending way. He was not accustomed to this treatment, and he could not let go, could not walk away from an ineffective performance before "his" people. Finally he got their attention by asking them—in what really was a demeaning manner for them and for him—to shout "Viva Kennedy" several times. Most of the crowd didn't grasp what he was doing, and didn't care. Finally he quit and left—through a squealing, pushing mob that still wanted to touch him, even if it wasn't particularly interested in listening to him.

On April 30, another unexpected development in a year of political surprises captured center stage. In Albany, Governor Rockefeller again changed his mind and said he would actively seek the Republican nomination after all. Johnson's withdrawal obviously had changed the picture for him too. Again insisting he would do nothing to split the party, Rockefeller said he would stay out of the primary states but would campaign actively elsewhere to prove he was the Republican who could win in November. A Rockefeller nomination probably would increase Kennedy's chances for the Democratic nomination, but Kennedy still saw the New York governor's candidacy as doomed, and he proceeded as in the past on the basis that Nixon would be the Republican nominee.

In Indiana, Kennedy had too much on his hands, with primary day only a week away, to be concerned about Rockefeller and Nixon. On the same day that Rockefeller announced his change of heart, Kennedy came to grips directly with the Pulliam papers. That morning the story about the McHale charges of "racism and religion" had appeared; at noon, when Kennedy went to Monument Circle in downtown Indianapolis and told a huge crowd he had been denied a permit to speak to them there, he quipped: "I don't want to spend my last five days in jail reading the Indianapolis *Star*." And that afternoon, addressing a group of clergymen at the Christian Seminary, he suddenly remarked: "I always thought the Manchester [N.H.] *Union Leader* was the worst paper in the country . . . I think, really, the Indianapolis *Star* must run it neck and neck. . . . I've been here two weeks, and I've never seen a worse paper. . . . It's certainly the most distorted, I think, one of the most warped. . . . Do they have one man who makes up all that? . . . It must be marvelous to have a toy like that."

The next day, the *Star* reported what Kennedy had said—after a fashion. The headline read, KENNEDY WON'T JOIN "POOR MARCH"—his casual response to a question about participation in the Poor People's March on Washington then being planned. Of Kennedy's blast against the *Star*, the story referred only parenthetically to "a digression which indicated his displeasure with some news coverage he encountered in Indiana."

But Kennedy had seen nothing yet. On the morning after he called the *Star* the "worst" newspaper, a month-long effort to convince its readers that Kennedy was "buying" the election reached its zenith. Up to then, the paper had been printing entirely legitimate stories discussing campaign expenses and quoting charges by State Democratic Chairman St. Angelo that Kennedy and McCarthy were spending $2 million in Indiana. Now the *Star* outdid itself by running on page one a reprint of a New York *Times* editorial under the headline IS INDIANA FOR SALE? ASKS THE NEW YORK TIMES. The editorial, questioning the spending of all candidates in Indiana, including Branigin, had distressed Kennedy when it appeared in the *Times* several days earlier. In fact, he had shown the poor taste of reprimanding *Times* reporter John Herbers, who had had nothing to do with the editorial, about it and about a campaign spending story Herbers did write, within earshot of fellow reporters at the Airport Holiday Inn. But the *Star* reprint was something else again. References in the original to Branigin's campaign and to Indiana's failure to have an effective law on reporting of campaign expenses had been deleted. The editorial as reprinted had a sentence that said, "Gov. Roger D. Branigin, the 'favorite son' candidate, is the leader of a state party organization." In the *Times*' original, that sentence had continued: "which controls thousands of patronage jobs and which still engages in the ancient and disreputable practice of levying a two per cent tax on the salaries of state employees."

Out-of-state reporters quickly ferreted out the subterfuge and wrote stories about the Indianapolis papers—if in fact they had not already done so. Roger Mudd of CBS and the Public Broadcasting Laboratory both reported on the Pulliam papers' coverage, and the attack on Kennedy finally subsided. It was cranked up again in the closing days of the campaign, more as a pro-Branigin effort than an anti-Kennedy one. Still, Salinger, seeing an exploitable issue in it, publicly demanded an investigation by the American Society of Newspaper Editors of the papers' "outrageous and callous disregard for fairness." It was, of course, too late for any such study by then, and was rejected out of hand anyway. The senior Pulliam responded by saying "Bobby Kennedy is like all spoiled children. When he doesn't get what he wants, he bellyaches about it. The facts are Kennedy and his entourage received more space in the Indianapolis *Star* and the Indianapolis *News* than any other candidate. . . ."

A postelection survey that appeared later in the *Columbia Journalism Review* showed, however, that in the nearly six weeks between Kennedy's filing in Indiana and the primary, Branigin received substantially more space in both papers, even when all the derogatory stories about Kennedy were included in his column-inch total. And, of course, it was not the sheer volume of space, but the editorial bias in the use of stories and cartoons that made the papers' performance so subject to criticism. Surprisingly, the *Star* printed many letters to the editor criticizing the anti-Kennedy slant. Also, it ran columns taking an opposing view—but not always happily. When columnist Joseph Kraft wrote that Indiana, "one of the last backwaters in the country, has finally entered the mainstream" and that it was bad news for Branigin, the *Star* ran the column but inserted this editor's note: "This article by Joseph Kraft, long-time columnist friend of the Kennedy family, ridiculing Indiana, is typical of the propaganda being turned out by pro-Kennedy writers to push the candidacy of Senator Robert F. Kennedy."

As primary day approached, Kennedy threw himself with seemingly boundless energy into his personal campaigning. If the electorate was going to decide on personalities, he was going to make the most of it. In a motorcade in the town of Mishawaka in northern Indiana, the crowd was so large and aggressive that Kennedy slipped, fell against a car, cut his lip and chipped part of his front tooth away. He had it capped immediately and went right on giving himself to the pawing mobs and trying to show he wasn't "ruthless" or impetuous. To one crowd he said, "Make like, not war. See how careful I am?"

On the final Sunday, Kennedy spent several hours back in the District of Columbia, which also was to vote the next Tuesday in a direct Kennedy-Humphrey test. The triumphal tour was saddened when a lead car struck the dog of a twelve-year-old girl. As the child stood numbly next to her pet at the sidewalk curb, Kennedy jumped from his car, stroked the animal and consoled the girl. Ethel ran into a store to phone for help. On the flight back to Indiana, Roosevelt Grier, the giant Los Angeles Rams football star, strummed his guitar and said of Kennedy: "This is the one the black man knows. He's the one who can do it."

In the history of American political campaigning, certainly in primary elections, Kennedy's final day of the Indiana campaign must be recorded among the most incredible. The Negro-blue collar base that Kennedy needed was there in Indiana, he and his strategists knew. And earlier trips through the Negro and potential backlash areas had persuaded them Kennedy could put the have-not coalition together. But getting it to the polls was another thing; lower economic neighborhoods tended to be the most apathetic about voting. Only if the residents were prodded, either by old-fashioned get-out-the-vote methods of physically transporting bodies to the polls, or by compelling interest in a candidate, would they vote in effective numbers. McCarthy voters were motivated to their eyeteeth: Branigin had the clout of the state organization to turn out the vote; Kennedy had whatever it was that turned crowds on.

With that fact in mind, Kennedy started in Evansville, the Democratic oasis in the conservative south, and flew to Fort Wayne, near the northeast corner. After a noon-hour rally that packed a downtown intersection, his motorcade headed for the airport and the next leg to South Bend. It was a warm, sunny day and the candidate was in a jovial mood on this final campaign tour of Indiana. He waved at small groups on the sidewalks as the car rolled at moderate speed out of the city. Suddenly Kennedy spied something and ordered the driver to stop. He jumped from the convertible and entered a small restaurant-bar bearing the improbable sign HUNGARIAN PIZZA. While aides bought the pizza and—of all things—apple strudel, the candidate walked to the small bar in the rear, shook hands with the proprietor, Zoltan Herman, and ordered a beer.

By this time, reporters were streaming from the press bus inside, buying up all the pizza and strudel in sight. Kennedy began passing beers back, and one led to another. Herman was a Hungarian freedom fighter who had saved a special bottle of wine for just such an occasion. He broke it out and handed Kennedy a glass. "To the Hungarian freedom fighters!" Kennedy toasted, and they drank. Herman refilled the glasses. Now it was the proprietor's turn. But he said nothing. "Who shall we toast now?" Kennedy asked, hopefully. No answer. "Does any particular name come to mind?" he asked. None did, although the bartender was drinking with a candidate for President of the United States.

Finally, Herman launched into a long only-in-America speech about the opportunities he had found here; more beers were poured and drunk as Zoltan droned on, apologizing at frequent intervals for his poor command of the language. "I *can* speak the language," Kennedy broke in at one point, "but when

I walk out of here I won't be able to speak at all." The minutes flew by, with Kennedy and the press party thoroughly enjoying the interlude—and Dutton and Barry nervously watching the timetable go to pieces. When they did persuade Kennedy he had to leave, the jam of newsmen quaffing beers around him made exit impossible. Whereupon the candidate, wine glass in hand, climbed onto the seat of an empty booth, thrust the glass out toward the press and began: "As George Bernard Shaw once said—" In seconds, the glasses were emptied, the trained seals rushed back to the press bus, and the motorcade was off, full of Hungarian pizza, strudel and beer.

After a short flight to South Bend and a courthouse rally at La Porte, there began what may have been the longest continuous motorcade in American political history. For nine straight hours, with crowds lining the road and city streets all the way, Kennedy stood on the back of a convertible through La Porte, Porter and Lake County shaking thousands and thousands of hands. In Lake County, there was no telling where one city ended and the next began. Sometimes there were so many reaching well-wishers that Kennedy simply put his arm out, letting it run along through the outstretched hands like Tom Sawyer scraping a stick along a picket fence. Kids ran alongside the car or rode their bikes for blocks at breakneck speed with seemingly endless endurance, shouting and waving. Michael and his sister were there again, calling to their friend. The crowd stretched from suburban road through tree-shaded residential section into downtown commercial street and out again, continuing along the open highway for long stretches, then back into suburbia, city residential and city commercial streets.

Early in the marathon motorcade, two young women in their twenties appeared along the route, waved at Kennedy, jumped up and down, then were seen to make a mad dash for their car. About a mile or so beyond, they were there again, going through the same routine. By actual count, the same girls appeared along the route thirteen times over the next few hours, just as enthusiastic each time. Finally some photographers invited them aboard their open car behind Kennedy, and they sat happily on the back for mile after mile.

Afternoon passed into early evening and then into dusk and darkness, and still the crowds lined the streets. A young boy with a basketball ran along for blocks, playing catch with the candidate when he called for the ball. Bill Barry, kneeling all the way and clutching Kennedy around the waist, eventually lost all feeling in his knees, which were bleeding by the time the night was over. After passing through Gary, the mayor of Hammond joined Kennedy in the car. Kennedy kept asking him, as block after block after block of humanity went by, "Where's the city line?" "Just another few minutes," he would be told. Kennedy later always referred to that motorcade as "Hammond, Hammond, Hammond." The schedule called for arrival in the town of Whiting for a last speech at 5 P.M.; it was after 10 o'clock when he got there. At one point, Kennedy passed a car that had a mattress on the roof. Three or four kids of pre-school age, dressed in pajamas and bundled in blankets, were stretched out on it. He stopped the motorcade and went back. "We tried to keep them awake," the mother said, "but they fell asleep waiting for you." The kids were roused and the candidate shook hands all around as he and Ethel held the smaller ones, sipping coffee brought out by the neighbors. (The press corps was eternally grateful to the kids for getting Kennedy to stop. Beer had been flowing for hours into the busbound newsmen; now they poured out, racing into the nearest bars, shops and private homes like men possessed.)

The crowd reaction was, in many ways, typical of what Kennedy had stirred in many other places around the country, though perhaps not at this pitch for so long a stretch in miles and hours. What set this motorcade apart, and what made it significant for Kennedy the candidate, was the unbroken display of adulation and support as he moved from Negro neighborhood to blue-collar ethnic back to Negro again, over and over and over. The neighborhoods ran smack against one another, and you could read their racial or ethnic composition in the faces that looked up at him, in the color of the hands that stretched out to him, in the accents that shouted at him, in the names on the simple storefronts. If the backlash whites were concerned about his strong appeal among the blacks, and his strong words in their behalf, there was little evidence of it that memorable night. Even after his last short rally in Whiting, as the motorcade moved toward Chicago, the crowds stretched for block after block. The campaign was supposed to end in Whiting, but he continued to reach for every hand that went out to him. "If they're waiting for me," he told the exhausted Barry, "we'll go see them."

The motorcade crossed into Illinois and raced to O'Hare Airport for the charter back to Indianapolis, where a closing reception long since had ended. Numb reporters staggered into the bar at the Airport Holiday Inn, where the Kennedy party stayed, for a nightcap. Two of them were sitting at a small table when the candidate came in, weary but obviously still keyed up after the incredible Lake County turnout, and strolled over to them. He didn't often come into the bars, and he declined an invitation to sit down. Instead, he stood at the table, looking back on the Indiana campaign and the final day. "Well," he said, "I've done all I could do. Maybe it's just not my time. But I've learned something from Indiana. The country is changing. . . ."

Kennedy went on like that, dwelling especially on the crisis in the cities, and on the need to reconcile the races. The old Democratic coalition of labor and the South is not the answer anymore, he said. Somehow whites and blacks in the northern cities had to be pulled together. He wasn't talking now in terms of a political power bloc, as he had with his strategists. He was expressing concern about the destructiveness to the country of racial conflict. More people in Washington—politicians, columnists—should get out around the country and see how worked up people were, he said.

Kennedy obviously had been greatly impressed and buoyed by the outpouring of sentiment for him on that last campaign day. Yet, as he looked back on it, he remembered a single derogatory sign held up to him along the way. It said, simply and bitterly, YOU PUNK. The man who held it, Kennedy recalled to the two reporters, ran ahead, turned and grabbed his hand as he came by on the back of the open convertible. "He squeezed my hand as if he were trying to break every bone in it," Kennedy said. The description, coming from the man who probably had shaken more hands than any other person alive, was offered with incredulity, that one man could have so much hate for another whom he did not know. One of Kennedy's listeners suggested to him, on the evidence of that day's motorcade, that he might be uniquely placed to preach and implement the message of racial reconciliation to both blacks and whites. Kennedy expressed hope that it was so, shook some more hands at neighboring tables, and left the bar.

Later, at dinner with some aides, friends and newsmen at Sam's Subway, an intentionally camp restaurant behind the Marott Hotel, Kennedy repeated the story of the man who had squeezed his hand. But he also dwelled on the kids who slept on top of the car, and Michael and his sister. He talked about the message of hope in the alert faces of the young poor, and how it faded as they got older. The people of Indiana, he said, had given him a fair hearing; they were tough, something like West Virginians in their independence. And he talked about the Pulliam papers and the spending stories. "I'd rather deal with the Indianapolis *Star*," he said at one point, "than with the New York *Times,* because you know where the Indianapolis *Star* stands." It was one of those pleasurable nights when he could stop being the candidate for a while. And then a man who had been drinking came by and made a crack about him. Kennedy just looked ahead, and said nothing. When the intruder had gone, he remarked, "You get so tired sometimes. You have to restrain yourself." And then, after that moment, he relaxed again and became a man having a late dinner in a restaurant with his friends.

On election day, Kennedy slept in at the Airport motel, played some touch football with the staff and reporters, walked Freckles and then relaxed in his room to watch the first returns on television. By 7:30 he was dressed and ready for his first night of victory as a Presidential candidate—or of defeat as any kind of candidate. O'Brien came in with the first sample precinct returns, and they were extremely good. In a lower-middle-class, racially mixed neighborhood, the vote was Kennedy 113, McCarthy 68, Branigin 46. In a Polish blue-collar precinct in South Bend, it was Kennedy 241, McCarthy 86, Branigin 62. And in a Negro precinct in Indianapolis, the count was Kennedy 341, Branigin 14, McCarthy 11! It was so good, in fact, that O'Brien asked his aide, Ira Kapenstein, to check out the figures. They were right. The word coming in from the District of Columbia was equally good. Kennedy's slate was trouncing two slates backing Humphrey. He eventually won 62.5 percent of the total there, to 37.5 for the Humphrey effort.

Kennedy and his wife went to the downtown headquarters hotel, the shabby Sheraton-Lincoln, where a host of friends and relatives waited to celebrate. Joan and Ted Kennedy were there, and Rene Carpenter, Susie Markham, John Frankenheimer, author Theodore H. White and a number of reporters.

The television networks, with their huge audiences, were clamoring for the candidates to grant special interviews. When the results became clear— Kennedy had won about 42 percent, to nearly 31 percent for Branigin and

27 percent for McCarthy—Kennedy went on with Roger Mudd of CBS. Because the CBS man with McCarthy, David Schoumacher, also had his candidate ready for an interview, a producer in the network remote truck in Indianapolis got the bright idea of switching the two candidates onto a split-screen arrangement. Kennedy, who had not wanted to give McCarthy undue exposure, was annoyed at the move, but answered questions from Walter Cronkite, the election-night anchor man. Cronkite suddenly asked Kennedy about campaign spending. Kennedy grabbed the opening. No candidate would have to spend so much money, he said, if the networks would give them air time at cost. With all the money the networks made, Kennedy said, they could well afford to do it, since the public owned the airwaves anyway. Cronkite obviously was sorry he had raised the subject at all.

"The only thing I wish I had thought to ask Cronkite," Kennedy told an aide later, "was what his annual salary was." The incident amused Kennedy because Cronkite had asked him the very same question about spending in a taped interview a few days earlier, for a special pre-election report on CBS. When the Senator gave the same answer, it was not shown on the air. Kennedy believed the exchange had been edited out by the network to save face, but Cronkite said later it had been cut purely for reasons of time, because the long interview included much more interesting material.

Kennedy was far from amused, though, about being linked with McCarthy in the split-screen technique. He told Mudd, for whom he had a high regard: "There are two ways you can play this—above the table and below the table, and your people played below. I had no indication this was coming. I want you to make it clear I feel this way." And he got up and left. Cronkite phoned Mudd shortly afterward and inquired whether Kennedy had been upset, but no answer really was required, it was so obvious. CBS was careful not to repeat the trick after that.

The television interview, one of several Kennedy gave that night, was not regarded as a laughing matter, either, by members of the writing press. Nor was the fact he appeared before his workers in the hotel ballroom immediately thereafter. The reporters had wanted the first crack at the candidate after his election victory, as had been the custom in the good old days before TV, and they complained to Mankiewicz. As a result, after his victory statement, Kennedy was ushered into the press room for a short news conference. Thus the pattern was set for election night in the future primaries—first, television appearances to reach the mass audiences, then down to the hotel ballroom for a victory speech to campaign workers, then to the press room for a news conference. One month and three election nights later, the same procedure was followed, with an outcome that shook the nation and the world.

But this night, Kennedy finished the press conference and returned to his hotel suite for some celebrating. The press session had been routine, except for the participation of an attractive blonde who asked, if Kennedy were elected President, "would you appoint your brother as Attorney General?" The candidate answered, "No, we tried that once." Then he asked the blonde: "Whom do you represent?" She wouldn't say. It was Joan Kennedy, wife of Teddy.

Now, up in the Kennedy suite, staff aides, family members, friends and newsmen crowded around in the tiny rooms, sipped drinks and congratulated the candidate, who was quietly pleased. Forty-two percent was not bad, what with the state organization against him, and at any rate, good enough to secure Indiana's 63 delegate votes for the first convention ballot. Beyond that, precinct breakdowns showed 9 of 10 Negroes voting for him in some areas and more than the usual number of blue-collar whites for a Democrat in the backlash neighborhoods. He had put the have-not coalition together for the base he needed. As the words to the parody had predicted, there were no ethnic problems on the Ruthless Cannonball. "I've proved I can really be a leader of a broad spectrum," he said to O'Brien that night. "I can be a bridge between blacks and whites without stepping back from my positions." What was more, the conservative approach and the personal campaigning in rural Indiana had paid off. He won 17 of 25 southern counties, 10 of 11 Congressional districts and 57 of the state's 92 counties. Looking ahead to rural Nebraska the next week, this latter outcome was encouraging. Finally, McCarthy had run third—though better than the polls had predicted. If only Kennedy could move him aside, get the committed college kids behind him again, he might be able to pull the whole thing off.

A reporter had brought into the suite a young girl named Mary Davis who had dropped out of college to be one of the early workers for McCarthy in New Hampshire, Wisconsin and now Indiana. She still was a firm McCarthy supporter, but was curious about Kennedy, and so the reporter introduced her—as a McCarthy worker who might be persuaded to defect (though it wasn't true). Kennedy, to Mary Davis' astonishment, literally growled at her. He was trying, humorously, to live up to his reputation in the McCarthy camp as a ruthless ogre. The girl was embarrassed, and Kennedy recognized it at once; he politely asked her where she went to school, and when she said she had dropped out to work for McCarthy, he said, "Good for you."

Later that night, when everybody had left Kennedy's suite and headed for the airport motel, Fred Dutton saw that nobody had remembered to bring Freckles, so he left Kennedy and went back for the dog. When Dutton arrived at the airport, he found his candidate sitting in the terminal coffee shop, deeply engaged in conversation with two other McCarthy college kids. Kennedy stayed with them in the coffee shop—the only place open then—until after five o'clock in the morning, talking. Much later, Dutton recalled about that night: "He had won in Indiana, but he couldn't win over those kids, and they really got to him. He had a lot of college kids with him, but he didn't have the super-activists, and he wanted and needed them, for himself. For days afterward, he talked about that boy and girl at the airport coffee shop—how great they were, in their idealism and determination."

But he didn't have them, and he didn't have time to brood much about it. In seven more days would come the second test—in the farm state of Nebraska, where Robert Kennedy of Boston and New York would be even more an outsider than he was in the prideful Hoosier state that somehow had seen fit to provide him his first essential stepping-stone to the White House.

CHAPTER FIVE

Nebraska

"DISPOSING" OF MCCARTHY

ON THE DAY after the Indiana primary, the Kennedy party flew back east to handle some other important political business before undertaking the short blitz campaign for Nebraska. The mood aboard the plane, not surprisingly, was light; champagne flowed and the candidate moved leisurely up and down the aisle talking politics and exchanging good-humored gags with aides and reporters. One newsman asked him, seriously, if he was going to change anything for the Nebraska campaign. "My last quote, if I can find a new one," he joked, referring to his George Bernard Shaw gem. Nobody ever had bothered to check it, but the "some men see things as they are" line really was an extremely free translation; in the original, far from being an exhortation for men to change life for the better it was part of the Serpent's temptation to Eve. For the poetic McCarthy it was a golden opportunity to take Kennedy down a peg, but he never did anything with it.

The important political business in the East was the rallying of New York Democrats to Kennedy's standard for the critical June 18 delegate primary election. Although the major candidates' names were not to be on the ballot in a Presidential preference contest, 123 of New York's 190 delegate votes to the national convention were at stake, with Kennedy, McCarthy and Humphrey delegate slates running in most of New York's 41 Congressional districts. Kennedy met more than 200 supporters at a lunch at the Inn of the Clock on UN Plaza and promised them he would campaign intensively in New York in the two weeks between the California primary and June 18.

The same night in New York, Kennedy spoke to the Synagogue Council of America and told its members: "Israel is not Vietnam. Indeed it is the very opposite of Vietnam. Israel's government is democratic, effective, free of corruption; its people are united in its support. We can and we should help Israel; but we can do this, I believe, secure in the knowledge that the Israelis—as they made so courageously clear during the June war—will not ask us to do their job for them; whether the terms of decision are idealistic or practical, helping Israel is in our most basic interest." In the midst of a hectic Presidential campaign, the endorsement was routine, and it received only passing notice. The issue of Arab-Israeli relations was not, after all, a particularly timely one, nor was it one of Kennedy's high priority themes; but he always had been a strong Israeli supporter, and the Jewish vote in New York was critical.

Before heading for Nebraska, Kennedy went to Atlantic City on May 9 to address the United Auto Workers convention. They had helped him in Indiana, and he would need them again, especially in California. In his speech, he dished out some strong criticism of the Vietnam War; the response was mild. Humphrey, who appeared before the convention later the same day, fared a little better, though not much. Organized labor would be a tough battleground for Kennedy against Humphrey, and he knew it.

But that would have to come later. The immediate field of combat was Nebraska, and the immediate objective there was to dispose of McCarthy. After his third-place finish in Indiana, many in the Kennedy camp regarded McCarthy's candidacy as a holding action for Humphrey. The Minnesotan had to be beaten emphatically enough in Nebraska to induce him to get out of the race, or to persuade his supporters to come over to Kennedy, so that the critical anti-Administration battle could be joined against the Vice President.

But McCarthy would not lie down. In Omaha on the day Kennedy was speaking in Atlantic City, McCarthy charged "the Kennedys" with "poisoning the well in Indiana." The remark was indicative of a growing bitterness on McCarthy's part. Everything had been going so well until the upstart New York Senator barreled into the picture. Indiana, McCarthy had told newsmen on primary night, really had been only "an interruption" in the campaign, because it was a three-man race that offered no clear test between him and Kennedy.

"I think the direct confrontation that was denied us here in Indiana," he had said, "will be given us in Nebraska."

McCarthy in Nebraska, however, acted as though he could not wait to be done with that primary. The campaign time was short enough as it was—just one week after Indiana—but McCarthy dawdled his way through a few days, then left Nebraska on the critical last weekend for the West Coast. In the process, he passed up the state's major Democratic event of the year, a Jefferson-Jackson fund-raiser at which Humphrey was to be the main speaker, with Kennedy and McCarthy also invited. It was strange conduct for a man who had said Nebraska would be the test he sought against Kennedy.

At the Blackstone Hotel in Omaha the night before McCarthy left the state, several reporters were having dinner when the Senator came over and sat with them. Nebraska wasn't very important, he told them in his customary off-handed way. Actually, he said, none of the primaries really mattered much, with the possible exception of California. "Kennedy talks about me winning three primaries," he said. "I've won five." The reporters knew of four he could claim having "won"—New Hampshire, Wisconsin, Massachusetts and Pennsylvania—but what was the fifth? "The college primary," McCarthy said seriously, referring to a recent nationwide poll of college students. "I'd rather win the college primary than the Indiana primary," he told his incredulous listeners.

Indiana, he proclaimed to them, had knocked Kennedy out of any chance for the nomination, presumably because he had not scored decisively enough. "All the cards are on the table now," McCarthy said. "It's just a question of playing them out." He himself didn't have many delegates to the convention, but that didn't matter either. He was the only Democrat who could get votes other Democrats could not, he insisted, and if the party wanted to win in November it would have to nominate him. Any Democrat would do well among Negroes and blue-collar whites against Nixon, he said; that wasn't the question. The independents, presumably the middle-class, better-educated suburban voters, would make the difference, and that was where he was strong. He predicted he would get 30 percent of the vote in Nebraska, Kennedy less than 50 percent.

It was clear from this conversation that McCarthy already was hedging against defeat in Nebraska, and that he intended to live to fight another day. Many of his key organization people had been sent from Indiana directly to Oregon and California. The fact that Nebraska was a farm state and he was supposed to do well there, coming from a Midwestern state and having served on the Agriculture Committee in the House, didn't seem to make any difference. As in the Indiana primary, McCarthy had challenged Kennedy to debate in Nebraska, but Kennedy had said any debate should also include Humphrey, and that wasn't in the cards. So, it seemed, there was little to keep McCarthy in Nebraska after that.

The Kennedy strategy for Nebraska hadn't counted on the possibility that McCarthy might cut and run. When Kennedy entered the race, he had no reason to expect that to happen. In fact, Nebraska had looked tough enough to the Kennedy advisers that it was decided to take a chance on Indiana a week earlier. Nebraska had been John Kennedy's worst state in 1960, and the Negro and blue-collar populations were relatively small. But Ted Sorensen, a native Nebraskan, had insisted Robert Kennedy could win there, so at the last minute he beat the filing deadline and was placed on the ballot. When Kennedy declared his candidacy in March, he had tried to gloss over his competition with McCarthy, noting that he was compelled by state laws to run in the Nebraska and Oregon primaries, both of which require the secretary of state to list all declared or likely Presidential prospects. It looked at that time that Nebraska might be the first primary Kennedy could enter without completely shattering the facade that he was running "in harmony" with McCarthy. That false front had collapsed, of course, in Indiana.

In any event, Kennedy was in Nebraska now, and he intended to make the most of it. The potency of the have-not coalition had been demonstrated in Indiana, and although the Negro population in Nebraska was less than 5 percent and the blue-collar population about 15 percent, together they figured to provide him a base, since nearly all voted Democratic. Again, getting out that base would be crucial, and here, under state campaign manager Jim Green, a prominent lawyer, and former Lieutenant Governor Philip Sorensen, Ted's brother, an effective organization was humming. Still, it was not certain that the Negro-blue collar base could be welded effectively in Nebraska, because Omaha recently had suffered riots, and backlash sentiment had been whipped up by a visit from George Wallace. But Kennedy would campaign hard in the have-not areas and hope for the best. And because strenuous personal campaigning seemed to have paid off in southern Indiana, there was hope it would work again in rural Nebraska. There were only a few days in which to go at it, but fortunately for Kennedy, the state's Democrats were concentrated in a tight arc running from Lincoln south and east, then north to Omaha, an easy day's motorcade. This would be the target area.

Although McCarthy was the prime opposition, it soon became clear that both Johnson and Humphrey would be factors too. The President's name remained on the printed ballot, his disavowal of candidacy coming too late to take it off. For the Vice President, a write-in effort of uncertain dimensions and cloudy paternity was started, in a state where write-ins were easy and popular. In 1964, more write-ins had been cast for Richard Nixon, Nelson Rockefeller and Henry Cabot Lodge in the Republican primary than there were votes for Barry Goldwater, the only candidate on the ballot. Humphrey's appearance at the Omaha dinner might spark a write-in spree for him, and Nebraska could become a muddled repeat of Indiana. But there was little Kennedy could do about that possibility—except campaign as he always did, as if there were no tomorrow.

On the morning of Friday, May 10, he left his Lincoln headquarters in a motorcade that took him to six southeastern cities and towns before reaching Omaha for the big Jefferson-Jackson Day dinner. Everywhere the crowds were large, warm and friendly. At the close of the first outdoor rally, in the town of Beatrice, the mayor presented Kennedy with a deed to an inch of land. To the crowd of housewives and downtown shop owners, Kennedy suggested that if he lost in Nebraska, maybe he'd settle in Beatrice. "And I'll bring Ethel and all eleven children," he added. The remark sent reporters scurrying to telephones; they recognized that the candidate had just confirmed a story about which his wife had refused to comment: that the Robert Kennedys were expecting again. Beatrice buzzed with the news all day.

There were, of course, the direct pitches to the special interests of the farmer. At the Otoe County courthouse in Nebraska City, Kennedy proposed a "Rural America Community Services Act"—a kind of rural renewal providing federal aid for development of social services to remote areas. It was skimpy, but it indicated the candidate's interest in the rural way of life. More than what Kennedy said, however, was the way in which he said it. He was loose, relaxed, cheerful with the crowds, and they were less frantic toward him, enthusiastic but well-mannered. When a middle-aged man began to heckle the candidate at the Crossroads Shopping Center in Bellevue in late afternoon, the crowd shushed him, and police led the man away. "Let him say what he wants to say," Kennedy called. "They've arrested me," the man shouted back. "Well," said Kennedy with an air of resignation, "if I'm elected President, one of the first things I'm going to do is get you out of jail." The crowd liked it, liked him; he wasn't so bad after all.

Perhaps in no other primary campaign did the private Kennedy surface quite so much—the quiet, humorous, even gentle man his friends knew and insisted to unbelieving outsiders he really was. Kennedy had won in Indiana, and the polls and his strategists were telling him he ought to win in Nebraska. Yet it wasn't cockiness; it seemed more that he felt welcome among these open and unsophisticated people, and relieved of the need to assault the Bad Bobby image.

That night, at the big Democratic fund-raiser, both Kennedy and Humphrey received enthusiastic—but noncommittal—receptions. Although Humphrey was the main speaker and Kennedy had been invited only to make remarks, the affair was billed in advance as the first confrontation of the two in the Presidential campaign. Kennedy seemed to be willing, and when he was scheduled to speak and leave before Humphrey arrived, he took his sweet time walking through the crowd, creating the impression he hoped to stall his departure into a direct meeting. As matters turned out, though, Humphrey did not arrive until after Kennedy had left.

Kennedy delivered his standard turn-this-country-around speech, full of youthful zeal and purpose. He came off somewhat like a law school debater before a benign audience of his elders. They were interested in seeing his style, but they were Democrats who had come to celebrate that fact and dream of future political glories; Kennedy instead gave them gloom and doom. Humphrey, by contrast, unleashed an old-fashioned political pep talk, belling the devilish Republicans at every turn. In one thinly veiled reference to the Kennedy road show, he said: "I do not think the American people want more frenzy. I do not think the American people will respond to leadership that exaggerates our difficulties to paralyze us with fear." The speech was a rouser, and the crowd enjoyed it, though the Vice President's listeners didn't seem especially convinced about anything. In the inevitable Kennedy-vs.-Humphrey comparison stories the next day, most writers thought Humphrey had "won" by a modest edge. What happened was that Humphrey had read the audience and its mood better than Kennedy did, and he delivered a speech better suited to the occasion. In any event, the speculation about a good write-in for the Vice President was not diminished any by the night's performances.

With McCarthy off to Oregon and California, the short Kennedy-Humphrey interlude shifted the next day to Humphrey's native South Dakota, where a Presidential primary was scheduled for June 4, the same day as the critical California test. There again, McCarthy and Kennedy were on the ballot and Humphrey wasn't, but could be the beneficiary of a large vote for President Johnson, whose name was listed.

Kennedy had decided late to go into South Dakota, but he had visited the state a number of times and was popular there. Also, a good organization, with Pat Lucey of Wisconsin revving it up, promised solid results. Some organized labor money was moving into South Dakota behind a Humphrey-Johnson effort, but the Kennedy organization felt it could win. Because of the importance of Oregon and particularly California, this probably would be Kennedy's last chance to campaign in South Dakota, so he squeezed it in, even though the primary in Nebraska was only three days off.

In the town of Aberdeen, Kennedy wasted no time unloading on native-son Humphrey for his "politics of happiness and joy." At a time American deaths in Vietnam were at a record high, he said, it was the wrong moment to paint rosy pictures. This pitch would come increasingly from now on, as the Senator tried to dismiss McCarthy and identify Humphrey as his major target.

In Huron, Humphrey's boyhood home, however, good taste and the time-honored idiocies of political campaigning required Kennedy to withhold his fire. Arriving shortly after the Vice President had paid a quick visit and left, he called on Mrs. Harriet Humphrey, Hubert's sister-in-law, in the Humphrey family drugstore. Walking arm in arm with her while townsfolk gaped, Kennedy picked out a Mother's Day box of candy and Mrs. Humphrey wrapped it, putting a Humphrey sticker on the outside. She refused any money, saying the candy was a gift from Hubert. It was all very homey and sportsmanlike. Then Kennedy returned to Omaha and reality, and a speech in a South Omaha park filled with harsh words of denunciation for the politics of happiness and joy.

Yet in this brief Nebraska campaign, Kennedy himself was exuding a glow of personal contentment that belied his words. He posed for pictures at the Sheraton-Fontenelle in Omaha with Mrs. Rose Kennedy on Mother's Day morning, went to church and had a late breakfast. The party flew to Lincoln and then motorcaded to the town of Wilber, where he helped dedicate "Czech Village," a low-cost housing project for the elderly. Striving always to find a link with his audience, he blurted out, "None of my children are Czech. But if things keep going on as they are now, one of them may be." (Pause.) "I don't know precisely what that means, but it brings us closer together."

In Crete, a metropolis of 3,546 souls at the last census, he told about a thousand of them: "You probably wonder why I came to Crete. When I was trying to make up my mind whether to run for President, I discussed it with my wife and she said I should, because then I would be able to get to Nebraska. So I asked her why I should get to Nebraska, and she said, 'Because then you might have the chance to visit Crete!'" The crowd clapped and cheered. "All those who believe that, raise your hands," he said. All the kids did, and their parents laughed. This particular bit of corn was a device Kennedy used over and over, starting in small-town Indiana, and the inherent ridicule in it seemed bad politics to city slickers in the campaign party. But there was no evidence that the majority of the grinning crowds ever thought so.

At the same stop, Kennedy noticed a sign for Governor Nelson Rockefeller. "Do you think Nelson Rockefeller understands the problems of the farmer?" asked this agrarian expert. "No!" the kids shouted back. "Does he care about Crete?" "No!" "When was the last time Nelson Rockefeller was in Crete? If he wants to be President, why doesn't *he* come to Crete?" Kennedy's own solution to the farmer's difficulties, he said, was buying twenty-six bottles of milk a day for his ten children. "So I'm trying to do something about the farm problem. I'm working on it right now. Let any of *them* match that by Tuesday." More cheers and applause.

And so it went through the warm, sunny Nebraska countryside. The next morning, the last day of the primary campaign, Kennedy stood on a platform in Wahoo, west of Omaha, pointed to a theater marquee and remarked, "I hope that's what you will make me tomorrow." The marquee read: THE HAPPIEST MILLIONAIRE.

In Wahoo, in the most casual, passing reference, the candidate told the crowd that "I'm very grateful for the fact you named your college after President Kennedy." And that was it. Although large road signs proclaimed Wahoo as the home of John F. Kennedy College and everybody in the entourage assumed there would be a big speech and rally there, the motorcade breezed straight out of town toward the next stop. It was baffling, and yet it began to dawn on many in the party that it had been a very long time since they had heard Robert Kennedy invoke the name of his brother in the old way. In those early days in Kansas and California, the references to John Kennedy, the New Frontier and the need to restore what had been were both blatant and frequent. At a stop for cold drinks along the highway, a reporter asked the candidate why he hadn't visited JFK College, and he replied, "My brother Teddy's been there." Fred Dutton confirmed the obvious. "We wouldn't have considered passing it up a couple of months ago. But it would have been too much now. He's a candidate and he wants to be a candidate on his own. Everybody knows he's President Kennedy's brother."

At the stop for sodas, a young girl from the Wahoo High School paper came up shyly and told Kennedy about a poll the paper had conducted. He had won with 127 votes, she said, to 61 for McCarthy. Did he have any comment? Kennedy thought it over a moment. "Thank those who voted for me," he said finally. And then, after a long pause: ". . . and tell the others I'll get them." The girl's jaw dropped, until he grinned, reached out and squeezed her hand. As the candidate stood there chatting, somebody remarked that it looked like a nice, easy day ahead. "Yes," Kennedy replied. "They've fired Marat/Sade as head of scheduling."

Just down the road was Father Flanagan's Boys Town, where a flock of teenagers awaited him on the broad green lawn. Accepting an honorary citizenship, Kennedy told them: "Actually, I worked my way up the hard way," and when he had them laughing, he talked to them as he always did before such audiences—earnestly, persuasively, although there was hardly a vote in the lot. "We have an election in the United States in 1968 that will have more effect on you than on me. . . . We've lived one half or two thirds of our lives, those of us running for President. And you have your whole lives ahead of you. There's an old saying, 'I complained because I had no shoes, until I met a man who had no feet.' " It was the happiest millionaire telling orphan boys they didn't know how well off they were, but somehow it didn't sound ludicrous; he had them with him.

The relaxed candidate breezed on into Omaha for a massive noon-hour rally at the Creighton University quadrangle. It was hot as he approached the microphones, so he peeled off his jacket and stood in his shirtsleeves. College, he told the students lounging on the grass and on ledges along the buildings, "gives you a license to remove yourself" from problems that plague the uneducated. "You can take it as a free passage . . . or you can figure a college education gives you a responsibility and an opportunity to involve yourself." The applause was sustained, and then he took questions.

What, other than money, was the difference between him and McCarthy? "Charm, sense of humor," Kennedy began. "I think he's occasionally ruthless, I don't know." Then, quickly, smiling: "I don't mean that. I don't want a headline, 'Kennedy Charges McCarthy . . .' I'm the one [who was ruthless]." After more playful banter, inevitably the subject turned to the draft. Rather than opposing it outright, Kennedy criticized its inequities, pointedly mentioning student deferments while poor and undereducated Negroes were being drafted. There were some restrained boos, and then a student asked: "But isn't the Army one way of getting young people out of the ghettos . . . and solving the ghetto problem?" Kennedy was incredulous.

"Here at a Catholic university," he asked, almost pleading, "how can you say that we can deal with the problems of the poor by sending them to Vietnam? There is a great moral force in the United States about the wrongs of the federal government and all the mistakes Lyndon Johnson has made, and how Congress has failed to pass legislation dealing with civil rights. And yet, when it comes down to you yourselves and your own individual lives, then you say students should be draft-deferred."

Kennedy asked those who favored the student deferment to raise their hands, and a majority went up. "How can you possibly say—" he began, agitated now. "Look around you. How many black faces do you see here, how many American Indians, how many Mexican-Americans?. . . . The fact is, if you look at any regiment or division of paratroopers in Vietnam, forty-five percent of them are black. How can you accept this? What I don't understand is that you don't even debate these things among yourselves. You're the most exclusive minority in the world. Are you just going to sit on your duffs and do nothing, or just carry signs and protest?" Again, he was at his gutsy best, telling it like it was, and most of them didn't much like it. In a short campaign week marked by bland, almost folksy patter and a lighthearted air, this Creighton confrontation was a memorable exception not soon to be forgotten by the students who were on the receiving end. For about twenty or thirty minutes, they saw Robert Kennedy with his emotions on his sleeve. Indifference, selfishness among the older, entrenched, undereducated citizenry, he could understand; when he found it among the supposedly aroused, activist youth, it appalled him.

The afternoon had become overcast when the candidate headed for Omaha's Negro section and a final speech there. The crowd bulged out of the Kennedy storefront headquarters and jammed the street, and as he spoke, the rain came pouring down. He continued to talk, with his head tucked under his raincoat, until it got so bad that he cried: "As George Bernard Shaw once said . . . Run for the buses!"

That night, instead of the customary closing reception at the headquarters hotel, Kennedy and his party flew to Columbus, Ohio, to deal with another pressing problem. Ohio Democrats held 115 delegate votes at the convention, and there was great pressure on the delegation to throw in with Humphrey. Such action would be a body blow to Kennedy's chances for the nomination. On request, the delegates had agreed to meet with him before reaching a decision. Determined to make a good impression, the Kennedy entourage for once arrived at the Columbus Airport on schedule. But the crowd there, and along his route as the motorcade moved through the city's Near East Side ghetto, was so huge that he arrived at the Neil House Hotel nearly two hours late. In the motorcade, however, were key members of the Ohio delegation, and they were greatly impressed by what they saw. After a subdued plea from Kennedy to remain uncommitted until after he could show his strength in the primaries, a majority of the delegates agreed. The state party's executive director, Eugene P. O'Grady, reported later that as many as eighty delegates had indicated, after having listened to Kennedy, that they would stay neutral a while longer. It was a grueling night, but well spent as far as the candidate was concerned.

On Nebraska primary day, time permitted no leisurely activities, as in Indiana. Kennedy flew to Davenport and Des Moines, Iowa, for two rallies and a meeting with that state's Democratic delegates. There were 30 delegate votes at stake, and, with the state convention approaching in ten days, a good chance that Kennedy would get the lion's share. The retiring Governor, Harold Hughes, was leaning his way; he often had seen the Kennedy pulling power and he was running in November for the seat long held by retiring Republican Senator Bourke Hickenlooper. "Harold Hughes will win in November," Kennedy told the Iowans, "because he is the right man for the job. . . . He has been a great Governor, and you and he together have made Iowa a model for this nation."

It was the stuff of which convention commitments are constructed, and the Kennedy of the old politics never permitted the Kennedy of the new politics to forget it. When he approached Larry O'Brien to join his campaign after Lyndon Johnson's withdrawal on March 31, Kennedy had assured the old politics' resident genius that he would do all the "right" things to enable O'Brien to perform his job. He was living up to that pledge; indeed, he would never have dreamed of doing otherwise. When you practiced politics, you made all the moves. Maybe McCarthy could campaign above the hoopla and could refuse to do business with the power brokers. But Kennedy understood and had been exposed to their clout too long and too closely to brush it aside. Harold Hughes was hardly the big-city boss, of course, but he was an establishment politician—one of the better ones, actually, with more spine than many others. It was easy for Kennedy to work with him.

Kennedy returned to Omaha from Des Moines by early evening, confident of a Nebraska victory. The NBC News preelection poll by Oliver Quayle had given him 47 percent, 22 percent for McCarthy, 7 percent for Johnson and only 9 percent for Humphrey on write-ins. With 8 percent undecided in that poll and 7 percent spread among minor candidates, Kennedy had reason to hope he might get a majority and dispose of McCarthy.

But at this stage, it was not clear at all how the Minnesotan was going to be moved aside. He showed no signs of caving in, and his growing bitterness and icy disdain of Kennedy made any direct approach virtually impossible. Time was growing shorter and the issue had to be drawn with Humphrey; perhaps if McCarthy himself wouldn't play ball, his supporters could be unglued from him by the reality that, if the Kennedy-McCarthy struggle continued, Humphrey would step over both of them and pick up the nomination. Especially in Oregon, the next primary in two weeks, it was important to shift McCarthy supporters into the Kennedy camp, because that primary was the one where the Administration's most energetic stop-Kennedy effort could be expected.

From the very first results that night, it was clear that Kennedy had done well. Salinger, in charge of producing obscure but persuasive voting statistics, manned the phones, bit hard on his ever-present cigar, and wrote down the good news from sample precincts to pass on to newsmen hard-pressed by deadlines. The Negro-blue collar base, centered around Omaha, had come through as hoped—about 85 percent for Kennedy among Negroes, nearly 60 percent among the blue-collar whites. In addition, Kennedy swept all counties with concentrations of European ethnics—Poles, Czechs, Germans, Scandinavians. And on top of that, the farmers turned out for him in droves. Of the state's 93 counties, Kennedy won 88, and the 5 he lost slipped from him by a total of about 50 votes. Of the 25 counties in which he waged a personal campaign, he won 24; the one he lost—the McCarthy stronghold of Lancaster, site of the University of Nebraska—got away by 2 votes. In all 12 towns with a population exceeding 7,000, Kennedy won. The final statewide figures eventually gave him 51.5 percent, McCarthy 31 percent, Johnson 5.6 percent, Humphrey, 8.4 percent on write-ins—a complete fizzle—and others, 3.5 percent. Kennedy had managed his majority, but again McCarthy had exceeded the estimates of the polls. He would not go away, and he would not play ball.

On television that night, McCarthy said he had gotten about what he had expected, and that he could have run about even with Kennedy had it been a head-on race. But it wasn't, he said, because of the Johnson and Humphrey votes, so he would have to go on to Oregon and California. "I don't think I have to win all [of the primaries]," McCarthy said. "Losing one or two doesn't make all that difference." He was out to raise the issues, he insisted, and the record still showed he could get votes that no other Democrat could.

Kennedy, in his television remarks, stressed his rural support in a conservative state and openly invited "those associated with Senator McCarthy" to switch to him. The directness of his pitch to them jolted many listeners and resurrected the "ruthless" image in many minds. But it had been determined he had to try to shake loose the McCarthy support in Oregon. Certainly McCarthy himself wasn't about to sign on. Of the Minnesotan, Kennedy said only, and without much conviction, "I would hope perhaps we could work together . . . in Oregon and California."

As in Indiana, the newspaper reporters were demanding a crack at the winner, so Kennedy came down to the Sheraton-Fontenelle press room. He declined to say he no longer considered McCarthy a serious candidate, but he noted he now had beaten him in Indiana and Nebraska, the latter a farm state, and the results were bound to have "an adverse effect on his candidacy." Kennedy said he had proved he was the popular choice, but "perhaps I'll have to go on and prove it again in Oregon, California and South Dakota." It would be a struggle to get the nomination "but if I win the three remaining primaries, I think it will indicate I have the support of the people across the country."

In the meantime, Kennedy made it clear he would be focusing on Humphrey from now on. The combined vote against the Administration in Nebraska—about 84 percent—"quite clearly demonstrates dissatisfaction with what we're doing at the moment," he said. "I think the people want to move in a different direction. We can't have the politics of happiness and joy when we have so many problems in our own country." As for McCarthy, he said, he had no plans to talk directly with him or to send an emissary to ask him to withdraw. "He's going to have to decide himself what he's going to do."

The next morning, Kennedy was off to Detroit for a speech and a meeting with Michigan delegates in advance of their June 1 state convention. As in Ohio, they were asked to stay neutral until after California. Before a crowd of 10,000 in Kennedy (formerly Cadillac) Square, the candidate teed off on Humphrey, and was cheered enthusiastically. "If you want to be filled with pablum and tranquilizers, then you should vote for some other candidate," he said, and he didn't have to be any more specific. Then it was on to Los Angeles, where he called on McCarthy backers to join him to prevent loss of the Democratic nomination to "the forces of the status quo." On Thursday, May 16, after an all-day swing between Los Angeles and Sacramento, Kennedy moved on to Oregon for the beginning of the next primary campaign—and the last before what was to be the critical test in California. It was three down—Indiana, the District of Columbia and Nebraska—and three to go, counting the South Dakota race on the same day as the California primary.

In Rapid City, South Dakota, that same Thursday, McCarthy spoke mechanically to a small crowd at the South Dakota School of Mines and Technology. "People say our campaign is failing," he remarked. "I'm not sure it is . . . but if we accomplish nothing more, it will be still regarded as the most successful and important political campaign, at least of this century." That was Gene McCarthy; he didn't know when he was licked, or wouldn't admit it.

CHAPTER SIX

Oregon

FIRST TASTE OF DEFEAT

RIGHT FROM THE beginning in Oregon, Robert Kennedy could tell it was different. His first speech there after the Nebraska victory, to the Portland City Club on May 17, was one of those throwaway appearances before businessmen that had to be done but didn't offer much for a Democratic candidate. He told them of a business magazine poll of 500 of their colleagues that turned up one Kennedy backer. "I'm the only candidate who can take all his business supporters to lunch," he quipped, and everybody laughed politely. When a businessman asked him about the government's role in birth control and he replied, "I don't want the federal government in *my* house," again the response was bland. But this wasn't his constituency, so the mild reception was no surprise.

Afterward, though, Kennedy went to Omark Industries, a chain-saw manufacturing plant where he might have expected to find some enthusiasm. Instead, the workers were restrained and cool. Then there was a stop at a nursing home, where the elderly gawked at this young man with the long hair, and then a visit to an antipoverty center in suburban Sunnyside where the crowd was so small and undemonstrative that Kennedy remarked, "I'm glad the police are here to keep them back." It was a poor day, any way you sliced it: poor scheduling, poor advancing, poor crowds. It ended with a mock convention at suburban Sunset High School, again a nonvoter exercise.

When Kennedy returned to the Benson Hotel he was ready to do some shaking up. There were only eleven days to go and he couldn't afford to waste any more of them. He picked up the phone and called for reinforcements—Larry

O'Brien, Steve Smith, Joe Dolan, John Nolan, John Douglas, others. O'Brien, who wasn't scheduled to go into Oregon at all but was to concentrate on New York, was shipped out posthaste. Smith and others occupied with California dropped everything and headed for Portland.

What had happened, Kennedy knew at once, was that he and the vaunted Kennedy juggernaut had miscalculated the challenge facing them in Oregon. McCarthy, after his Indiana and Nebraska defeats, was supposed to fold. But he wouldn't. Now, in Oregon, it was tough, and the Kennedy strategists didn't even have the first team in. Their experienced manpower stretched thin by the punishing schedule of six primaries in four weeks and with critical California coming up, they had relied on local evaluations and tried to get by with second-string organizing talent. "It was somewhat of a stepchild," Steve Smith said later, "and when you recognized the problem it was almost too late."

The candidate's late entry into the Presidential race hurt him in Oregon particularly; in giving McCarthy a head start in a state already strongly anti-Vietnam, Kennedy had lost a shot at important Democrats who might have been with him. So he had turned to the state's most influential woman in politics, Representative Edith Green, to provide local stature. Mrs. Green was a good campaigner and an able legislator, but bigtime national politics was not her forte. Yet she was strong-willed and would not settle for being just a figurehead. The out-of-state Kennedy people to be sent in had to be cleared with her, and she got men who had neither the experience nor the organizational ability to cope with this kind of sticky challenge. The first was E. Barrett Prettyman, Jr., a Washington lawyer with some business connections in Oregon—and Mrs. Green's confidence. On April 29, vanden Heuvel was sent in. He had worked on the 1960 campaign, lost a Congressional race to Republican newcomer John V. Lindsay and labored on the fringes of New York politics as a Kennedy ally. A third import was Herb Schmerz, an able advance man. But they were in over their heads, and now would have to be bailed out. "One thing about the Kennedys," said one of the candidate's intimates who was at the Benson that night, "when they get into a situation like that, they don't want discussion. They don't want to hear people around the room saying, 'Maybe we should try this or maybe we should do that.' If the thing is really bad, they know what they want done and they just order it done. That's what Bob did."

The new team quickly recognized the weaknesses: first and foremost, the all-important Kennedy base—the have-not coalition—was not there. In one meeting, when a lately arrived Kennedy strategist asked, "Have we got the ghettos organized?" Mrs. Green, a bit indignant, replied: "There are no ghettos in Oregon." And she was just about right. The Negro population was a tiny one percent and what ethnics there were had long ago been assimilated into this affluent, comfortable northwest paradise. "This state is like one giant suburb," Kennedy himself said, and that wasn't far wrong either.

Except for the Kennedy message on Vietnam, the Senator's recitations of a troubled America did not square with the personal experience of most Oregonians; they didn't see it anywhere around them. And on the war, somebody else—Gene McCarthy—had gotten to Oregon well ahead of him. A state to which the Vietnam War debate had come much earlier than to most others, Oregon was conditioned to the acceptance of a peace candidate by the anti-Vietnam views of its two Senators, Democrat Wayne Morse and Republican Mark Hatfield. When McCarthy announced his candidacy on the last day of November, liberal Oregonians already were in the vanguard of the dump-Johnson movement. Not simply college dropouts, but concerned adults with political experience, joined in to give McCarthy an organization more solid than he had in any of the earlier primary states.

Also, Oregonians had a reputation for independence, and for taking their primary seriously. It had been the first "free for all" election in the United States requiring the listing of all prospective candidates; the voters did not have to be told, as in Indiana, that they had a chance to pick the next President. They also were, it was said, more issue-oriented than voters in many other states, and less swayed by campaign hoopla—by charisma and glamour. And finally, Oregon was a state in which organized labor had unpleasant memories of Robert Kennedy. As a Senate rackets investigator he had focused national attention on Portland's vice and labor corruption and its mayor, Terry Schrunk. In Oregon, they knew the difference between John Kennedy, whom the voters endorsed in the 1960 primary, and his brother Robert.

At one school rally, a young man, seemingly gone berserk, charged after Kennedy as he was leaving. Dick Harwood of the Washington *Post* wrestled him down and held him until he promised to cause no trouble. But when Harwood let him up, the youth pushed through the crowds determined to get to the candidate. Barry, already alert to the threat, intercepted him, put him to the floor and got spectators to sit on him until authorities could be found and Kennedy could make his exit.

Beyond the unfavorable feeling toward Kennedy, Oregon was ideally suited to McCarthy. It was middle-class, economically comfortable, relatively well-educated. Like Kennedy, its people rejected the war and Lyndon Johnson, but many preferred McCarthy's quiet, thinking man's approach. In all the primaries, McCarthy had done well in the suburbs, and if Kennedy's assessment of Oregon as "one giant suburb" was right, it was made to order for the Minnesotan.

So what it all got down to was this: Kennedy, not having his usual base, would have to be more effective than ever as a personal campaigner. In the remaining days, his staff would have to get him around the state and hope he could break through in this statewide suburbia. But he should not come on too strongly either; this was Oregon. Nor could it be forgotten that looming ahead, just one week after the Oregon May 28 vote, was big casino: California and its 174 delegate votes.

On Saturday, May 18, the candidate staged another whistle-stop, this time on the Southern Pacific's Beaver State Special through the Willamette Valley, the vote-fertile area running south of Portland about 115 miles to Eugene. The back platform of the six-car train was decked out in red, white and blue, and at each stop the candidate spoke to crowds that ranged from just moderate at first to good as the day wore on. As he moved away from Portland, he appeared to be doing better, but it was hard to tell because as usual there was an abundance of screaming kids in the trainside turnouts. Their mood rubbed off on him, leading him to make the playful remarks that amused some and cooled others. Of the importance of Oregon to him, he said, "Can you imagine the conversation with my children at home? They'll say, 'Daddy, you did well in Indiana and Nebraska, but how did it go in Oregon?' If I have to tell them I lost, can't you see the tears coming to their eyes and running down their little cheeks? You wouldn't want to do that, would you?"

The overall strategy, however, was clear: continue to ignore McCarthy, focus on Humphrey and the Administration, and shake every hand in sight.

He jumped on a Humphrey statement—and later "clarification"—indicating there had been agreement at the Vietnam talks in Paris to include the National Liberation Front; "if such confusion can be created among members of the American government," Kennedy argued, "agreement with other parties will be difficult indeed." And in a blatant pitch for the women's vote, he called for appointment of a woman to the American negotiating team. "They're the ones who have to give their husbands," he said. "They're the ones who have to send their sons. They're the ones that are making such a great sacrifice in addition to the men in Vietnam."

After stops at Oregon City, Salem, Albany, Harrisburg and Junction City, the Kennedy Special pulled into Eugene, home of the University of Oregon and a McCarthy stronghold. The crowds were encouraging. They lined the downtown streets as his motorcade pulled through—past the tell-tale signs of advance work: ROBERT KENNEDY WILL PASS THIS POINT AT 1 P.M. (The signs always gave an arrival time that was too early, to allow crowds to gather.) At Hayward Field, an old wooden stadium, about 6,000 spectators turned out. There was a small circus encamped on a field nearby; the candidate thanked the crowd for electing to come to see him instead. He called the city "the Eugene I like best in the country," a remark that generated some boos from the McCarthyites in the audience. Generally, though, his acceptance was better than it had been on that dismal previous day in Portland.

But everything is relative, as the rest of Saturday clearly established. Because of the importance of California, it had been decided to squeeze into the schedule a three-day swing south, the plane arriving in San Francisco early enough that night for a downtown motorcade and a major speech to northern California party workers. As the candidate moved toward South San Francisco, through a large shopping center and on to San Francisco's Mission Street, the contrast was ominously striking. The area was heavily Mexican-American, a key element in the Kennedy base in California, and the outpouring of enthusiasm only compounded the inner-circle apprehension about Oregon. There were in San Francisco, to be sure, many McCarthy signs, including taunts about Kennedy's refusal to debate, but they were washed away in a sea of adulation.

At such moments, Dick Drayne was at his best as busmaster. Along Mission Street, when a strange car cut into the motorcade ahead of the press bus,

Drayne jumped off the bus, ran ahead, planted himself in front of the interloper and waved the car out of line to the left—into a stream of oncoming traffic. The dumbfounded driver obeyed. What happened to him thereafter remained one of the minor untold stories of the Kennedy campaign.

That night, at the Sheraton-Palace Hotel, as if regenerated by the California response, Kennedy gave one of his most confident, professional performances. He didn't say anything he hadn't said in Oregon; but it seemed to come with more conviction, and again, as in the early days in Kansas, the crowd enthusiasm appeared to create an emotional playback on the candidate himself. He did not seem merely to be talking; he was hooked on his own message of reconciliation between black and white, have and have-not. It gave all the frantic campaigning, and his own public existence, a purpose of their own. If only he could get by bland Oregon, he would be home free.

The party went to Los Angeles that night, and although Kennedy might have been expected to be jubilant over the San Francisco triumph, his mind was on Oregon. He roamed up and down the aisle of the chartered jet, talking about the contrast with California and about the problem of dealing with an electorate that had no troubles. "Let's face it," he told James Dickenson of the weekly *National Observer* that day, "I appeal best to people who have problems."

The proof of that observation, if any were needed by this time, came on the next campaign outing. Kennedy took Sunday off, a rare thing for him, at John Frankenheimer's beach house in Malibu. But Frankenheimer kept a flow of his own social friends moving through and Kennedy finally sought refuge in a restaurant down the coast with Goodwin. On Monday, he set out on one of those marathon motorcades that exhausted the rest of the traveling party and reduced Bill Barry to numbness. The candidate first toured the San Gabriel Valley, flew south to San Diego and then returned to Long Beach for the start of a fifteen-mile, three-hour ordeal through the jammed streets of East Los Angeles. Raucous crowds, heavily Mexican-American, in their enthusiasm nearly pulled him from the back of the convertible four different times. Nuns wore "Kennedy" bumper stickers across their starched cornets. As the warm afternoon dragged on, Drayne in his capacity as keeper of the press would dash periodically from the motorcade into a bar or grocery and return with cold beer for the troops. At one juncture, a white-haired columnist not yet approaching senility protested that not everybody drank beer. "I tried to get you a six-pack of Geritol," Drayne replied, "but they didn't have any."

Meanwhile, up ahead of the press bus, Kennedy looked down from his perch and saw a young boy busily removing one of the candidate's shoes. Before Kennedy realized what was happening, the boy had taken the other one too. (He wore them to school the next day, an enterprising Los Angeles *Times* reporter found out later.) Kennedy just stood there, in his stockinged feet, shaking hand after hand until the motorcade reached its destination, the Temple Isaiah in Los Angeles, well after dark. As they were about to go in for a speech, Kennedy looked down. "My shoes," he said, turning to Dutton. "I'll have to borrow yours." So the faithful lieutenant took his off and waited in the car. "I'd have been here sooner," Kennedy quipped to his long-waiting audience, "but Mayor Yorty [his archfoe] gave me a police escort. A while ago we passed a sign that said: 'Buffalo, 175 miles.'" After Kennedy's talk at the Temple, the candidate directed his driver to pull out of the motorcade and head for a favorite swank restaurant. Kennedy walked in and Dutton, shoeless, followed. (Four nights later at a huge fund-raising gala in Los Angeles, Kennedy began his speech by saying he was indebted to his father for steering him into public life, to his wife for her encouragement along the way, and "to Fred Dutton for my black shoes." It was a typical Kennedy throwaway line, that only Fred and a few newsmen understood.)

On page one of the Los Angeles *Times* the next day, the Kennedy campaign story featured his remarks at the Temple: "Capping a hectic day of campaigning in Southern California [the story said], Sen. Robert F. Kennedy (D-N.Y.) Monday proposed a U.S.-Soviet agreement to stop the huge arms shipments to warring Jews and Arabs in the Middle East. . . . He emphasized, however, that he feels that until such an agreement was reached, the United States 'must fully assist Israel—with arms if necessary'—so long as Russia continues sending arms to the United Arab Republic and other countries hostile to Israel." It was a routine account of a routine speech, one that most readers doubtless only skimmed over.

Kennedy flew up to San Francisco on Tuesday, May 21, for a speech before the local press club. In the Q-and-A period, he was asked what would happen if he lost in Oregon. "If I get beaten in a primary," he replied with more candor than political dexterity, "I'm not a viable candidate. I might be a nice man . . . I'd return to being unruthless if I lose in Oregon." It was one of those remarks an otherwise careful politician makes that leaves his strategists talking to themselves. Not that the premise wasn't entirely valid; he was, after all, pinning his whole campaign on the idea that he was clearly the people's choice. It was just that you didn't, if you were savvy, talk yourself into a corner from which there was no diplomatic escape.

That lesson had been learned early by McCarthy. After his Indiana defeat he had said Nebraska would provide the real confrontation with Kennedy. But after he had lost that primary too, he brushed it aside as insignificant. In Oregon, when reporters asked McCarthy whether the crucial test finally was at hand, he told them irritably, "I wish you people would stop talking about crucial primaries. Whether you win or lose a primary or two isn't that important. Humphrey hasn't won any primary but people are calling him the front-runner."

With Kennedy, though, it was different. The one thing Kennedys were supposed to do, no matter what else you thought about them, was win. McCarthy, in the minds of all the pros, never really had been "viable" anyway. He was regarded as a man fighting for a cause, not for the White House. That was the general viewpoint, at least, before New Hampshire and Wisconsin. Then he seemed to see himself more as a bona fide candidate. But when Kennedy abruptly interrupted his reveries in Indiana and Nebraska, McCarthy's cause—ending the fighting in Vietnam and challenging the politics of the Cold War—seemed to be shunted aside. What appeared to take precedence was a growing personal vendetta against Kennedy. It revealed itself in both private and public remarks by McCarthy, by pettiness and bitterness. Part of it, some of his aides insisted, was a matter of pride; Kennedy simply was ignoring McCarthy, and it was getting to him. Demands for public debate were being dismissed out of hand. And Kennedy was getting away with it, as the primary results showed. His highly emotional base—the blacks and ethnics—were tuned out to the unemotional campaign of reason McCarthy was waging, and tuned in to the Kennedy charisma and campaign of restoration. Or so it looked from the McCarthy camp.

Now, in Oregon, McCarthy realized as well as Kennedy that the have-not coalition was not present in force. He began to draw attention to the fact that

Kennedy's base in other states had come from voters of lower education—and to the fact that the Oregon electorate was an elite one. "Oregon has the reputation of passing a most enlightened judgment on candidates," he told Oregonians. "I was quite willing to move along, believing that after Wisconsin, the real test of my candidacy would come in Oregon." And to another audience, he said: "The polls seem to prove that he [Kennedy] is running ahead of me among the less intelligent and less well-educated voters of the country. On that basis, I don't think we're going to have to apologize or explain away the results in the state of Oregon." To students at the Oregon State University at Corvallis, McCarthy discoursed on the quality of the Indiana electorate, ridiculing Hoosierism with such acidity that it caused even the most loyal McCarthyites to wince. "They keep talking about 'the poet' out there," he said. "I asked if they were talking about Shakespeare or even my friend Robert Lowell. But it was James Whitcomb Riley. You could hardly expect to win under those conditions."

In this atmosphere of disdain, McCarthy granted an interview to Jack Cole of Metromedia Television News on a plane trip from Klamath Falls to Coos Bay on the same day Kennedy was committing his "viable" blooper. Asked whether, if he were eliminated in the remaining primaries, he might lean to Humphrey or Kennedy for the Democratic nomination, McCarthy indicated he conceivably could back Humphrey if the Vice President were to change his position on Vietnam. The statement, in the context of McCarthy's all-too-apparent feelings about Kennedy, sent temporary shock waves through the McCarthy camp. The younger people especially, who didn't like Kennedy either but to whom Humphrey represented the war and the Johnson establishment, were shaken. McCarthy at first insisted he had been misquoted—though his words were on film—and then he "clarified" his statement. He was, he said, in a position of "absolute neutrality" between Kennedy and Humphrey, and was still fighting both of them. He "could" support Kennedy if the New York Senator conducted himself properly. This last was a reference to still another source of personal McCarthy irritation—the "distortion" by Kennedy backers, McCarthy said, of his Senate voting record. But no clarification could remove the growing impression that what McCarthy wanted to do more than anything else was to even the score with Kennedy for coming in and muddying the waters.

Kennedy, of course, had asked for it by the manner of his entry into the campaign. His hope of teaming with McCarthy to fight the Administration was at best a self-delusion and at worst an early tactical facade. Now it was clear the focus of Oregon was becoming stop-Kennedy—to prove, in Kennedy's own words, that he was not a "viable" candidate.

In this effort, McCarthy was getting considerable help from the independent efforts of the Administration's backers in organized labor. The AFL-CIO's Committee on Political Education (COPE) launched an intensive phone campaign urging 50,000 Oregon union members to vote for Johnson, whose name remained on the ballot. Women volunteers in the Portland Labor Center were manning a battery of twenty phones for thirteen hours a day, with other volunteers doing the same around the state. All had the same message, printed on a card from which the volunteers read: "Good morning. This is Mrs. ——— calling from the COPE office urging you to vote for President Johnson and Vice President Humphrey in the primary next Tuesday to show our support for them in the peace talks in Paris. Give them your hand. Thank you."

This was the squeeze on Kennedy as he returned to the Oregon campaign trail on May 22 for his last intensive six-day push. An early NBC News poll had given Kennedy the slightest edge over McCarthy, 29 to 27 percent, with Johnson getting 16 percent, Humphrey 12 percent on write-ins, others 4 percent, and undecided 12 percent. But still the decision was to ignore McCarthy, hopefully cooling the "ruthless" problem, and to focus on Humphrey as the heir apparent of Lyndon Johnson.

From Pendleton in the northeastern corner of the state, Kennedy undertook a fifteen-hour barnstorming swing through remote eastern Oregon, discussing farm and irrigation problems, and Hubert Humphrey. Some of the airports were so small that it was necessary for the party to use two small planes, a Convair and a converted C-54. To get out at La Grande, Kennedy had to climb aboard a raised forklift truck. To a large crowd at Eastern Oregon College, he said: "If I do badly in Oregon, the Vice President is going to be the major gainer. That's why his people, in his behalf, are having a telephone campaign in Oregon at the present time—to bring out the votes to see if he can defeat me here and also in California. He is the major opponent. I am running against the period of the last three years. As the Vice President came into Nebraska [for the Omaha fund-raiser], it would be well if he would come into the state of Oregon and answer questions as I have answered questions." The words echoed the problem as summed up a few days later by Mrs. Green. "We're running against a ghost," she said. "The money is coming in, but the body isn't." That, of course, wasn't quite right. Humphrey wasn't there but McCarthy was, and he was proving to be no ghost.

Kennedy drew some hope from eastern Oregon. The crowds were warm and friendly, like Nebraska. In The Dalles, a tiny place where he actually used a stepladder to disembark from a DC-4, a crowd of 3,000 lined the streets, apparently some sort of record. A local radio commentator in a mobile unit in the motorcade waxed so ecstatic that the reporters on the press bus tuned in on portable radios to marvel at his description. "My God, there must be two hundred people at this intersection!" he would say. Dick Tuck, the campaign jester, ran back to the mobile unit, pouring forth superlatives about the turnout, which were reported at once to The Dalles' vast radio audience.

But eastern Oregon was not going to make much difference; the Portland area continued to be the key, and the problem. The next morning, Kennedy took a sightseeing boat onto the Willamette River, to hear about harbor problems. It was another of those dead spots in the campaign, where no voters were met or talked to, except a few local bureaucrats. Kennedy was gloomy and it was written all over his face. As the party was leaving the boat and walking up the pier. I asked him how Oregon was going. "I've got a problem here," he conceded without hesitation. Well, what was it? "You're a political writer," he told me quietly. "You can look around and see what it is." And he declined to say more. "I don't want to play games. I have my own analysis, but I don't think it would be useful to me to go into it." What he meant, of course, was that he did better "with people who have problems."

Later that same day, Kennedy had more problems of his own. He got word that columnist Drew Pearson had written a column for the next day charging that Kennedy, as Attorney General in 1963, had ordered a wiretap placed on the late Dr. King. The allegation, in a column that regularly appeared in four Oregon newspapers, held that the wiretaps had produced reports on a "Communist who was helping to write King's speeches," and also reports dealing "in part with his [King's] sex life." Many papers around the country, including the afternoon *Oregon Journal* in Portland, ran the column but deleted the latter reference, which went on to say that an FBI report quoted "a confidential informer as claiming that Dr. King has been having an illicit love affair with the wife of a prominent Negro dentist in Los Angeles since 1962." Pierre Salinger obtained a copy of the Pearson column prior to publication and said flatly that Kennedy, as Attorney General, "never authorized eavesdropping of any kind." He noted that Kennedy "did authorize wiretapping in cases where national security was involved, and where there was a written request from the FBI." But he added that Kennedy "has never commented in the past on individual cases, and he will not so comment now." The question was, of course, not eavesdropping, but wiretapping, and when Salinger was pressed to say whether Kennedy had requested the specific wiretap to which Pearson referred, the answer again was noncommittal: "Senator Kennedy has never discussed individual cases and isn't going to now."

In the scheme of things, the story didn't have much to do with Oregon politics or the primary. Kennedy aides were quick to point out, though, that "Mr. Pearson is a declared supporter of another Presidential candidate and he in fact had a fund-raising dinner in his home for another candidate [meaning Humphrey]." It was a tempest in a teacup, but in a campaign in which the ruthlessness tag on Kennedy again had become a factor, it was still another headache.

At this stage in other campaigns, everything had begun to go right. In Oregon it was just the opposite. The next morning, Friday, May 24, Kennedy made a tour of Oregon's coastline. At Astoria, walking barefoot along the beach with Ethel and Freckles, the candidate suddenly had an urge to take a dip in the 54-degree waters. Obtaining a no-pictures pledge from the accompanying photographers, he and Ethel walked on another hundred yards or so, where he stripped to his shorts and plunged in alone. No pictures were taken, but words were written, and in the opinion of Oregon political seers later, the episode probably cost him votes. Candidates for President, it seemed, were not supposed to take off their clothes on the beach and dive in—not in Oregon anyway.

While Kennedy was thus occupying himself, McCarthy was releasing to the press a telegram informing Kennedy that he had bought thirty minutes of television time in Portland for a debate. "Campaign rallies and paid advertisements," McCarthy intoned, "do not meet our obligation of conscience to give the people a fair opportunity to compare our positions."

Accepting TV time that McCarthy had bought was, tactically, out of the question. But the matter of a debate, for which free time certainly could be obtained, now required serious reconsideration. Kennedy had ignored McCarthy, but he was still there—very much there.

The younger members of the Kennedy staff, notably Walinsky and Greenfield, felt that the time had come for their man to take McCarthy on directly. They were confident that in debate their boss could establish that he was more thoroughly grounded in the issues than the seemingly lackadaisical McCarthy, and that such a confrontation could do for Robert Kennedy what John Kennedy's first 1960 debate with Richard Nixon had done for him.

It was Saturday, May 25, just four days from the primary, and the morning had been another of those bad ones: a street rally scheduled for the wrong hour, and in the wrong place, next to a carnival whose ferris wheel nearly drowned out the candidate's voice. Kennedy returned to his eighth-floor suite at the Benson, chagrined, and held court with his aides on the possibility of debating McCarthy. The standing argument against it had been that such a televised meeting would only give the Minnesotan more visibility. But ducking a challenge was not in the Kennedy style. Still, the object was to win, not to prove personal courage. Kennedy had asked Mrs. Green about it, and she had said, flatly, "No." Walinsky, Salinger, Pat Lucey and vanden Heuvel came in and all said they thought he should debate. Kennedy himself was not persuaded. The polls showed him slightly ahead and he still hoped he could pull it out; in his mind the debate wouldn't make that much difference that late in the game.

But Walinsky and the others continued to press. This kind of persistence, from his young Senate staffers especially, annoyed Kennedy. Over the last few years, they had exerted a profound influence on his thinking and he valued them as a link to the new politics and to the young. But when it came to hard-nosed tactics, he saw them as they were—inexperienced, excitable, compulsive. Just as he bypassed them for the old-timers in his critical explorations about the wisdom of running at all (about which, in retrospect, the young aides were right), he was impatient now with their efforts to exert tactical influence, and he turned instead to the old hands. A political campaign, in the Kennedy view, was not an intellectual exercise, especially when you were staring down the jaws of defeat. O'Brien was called in, and Kennedy laid out his pragmatic reasons for not wanting to debate: it gave McCarthy exposure, it smacked of desperation, it could take away a needed weapon in California. "Wherever we are on Saturday, we're going to be on Tuesday," he said. O'Brien, who acknowledged that his candidate was in trouble, indicated he didn't think it would hurt to debate, but would go along with Kennedy's decision. That seemed to settle the matter.

It was now late in the afternoon, but the pressure from Walinsky, Greenfield and the others persisted. Kennedy called Mrs. Green again. Again she advised against it. There would be no debate. Kennedy retired for a nap before dressing for the evening's rally.

But the debate now had taken on a make-or-break importance in the minds of its proponents. They went to O'Brien. His feeling still was that it wouldn't have much impact either way; if Kennedy wanted to do it or not, it was okay with him. The proponents, in a high state of agitation, went to Dutton's room down the hall. The matter, Walinsky insisted, had to be reopened with Kennedy. Reluctantly, Dutton went down with them and got the candidate up. It now was past six; they stood in the living room, still pressing their demands. Kennedy, standing in their midst in his shorts, was adamant: the decision had been made. But they continued to balk. Kennedy ordered the room cleared except for a few of the old professionals. He called one of them into the bedroom. "They're pressing in on me," Kennedy told him. And then, looking at the old pro, he added: "Don't tell me you're buying these guys." It was a political crunch, and time for political decision making. "I don't know what they're doing here," Kennedy said of his young speechwriters. "I didn't even want them out here."

Outside in the corridor, Walinsky, Greenfield and others were milling around, talking loudly. Kennedy, still dressed only in his shorts, went to the door, opened it and stalked out. "I thought we decided that," he told them angrily. "Why are you standing around here making noise? If you want to do something, go out and ring doorbells." And then, turning to Adam and Jeff, he barked: "Besides, I don't see why my speechwriters aren't writing speeches instead of playing the guitar all the time." And he stormed back inside. This last was a slow burn erupting. Sometimes during the campaign, Walinsky and

Greenfield would get aboard the campaign plane and start playing and singing folk songs while others, including the candidate, tried to work. Now the group in the corridor broke up, and Kennedy dressed for the rally. Adam and Jeff came to Dutton's room later, properly chastised, announcing they were returning to Washington to write their speeches. Dutton laughed it off and told them to do the same. That was the end of it—but there was no debate in Oregon.

That night, while Kennedy attended a routine high school rally, McCarthy was drawing a tremendous and wildly enthusiastic crowd at Portland's Memorial Coliseum. After a brilliant and rousing recounting of the force behind his own candidacy and what it already had achieved, McCarthy unloaded on Kennedy and his "special constituency," openly and in a style more befitting him than earlier private carping.

The New York Senator, he said, "has brought some confusion of the old issues . . . but little clarification. He's not brought any New Politics to the scene in 1968. We had something going and still have in New Politics in this year, but it doesn't consist of adding up somehow a consensus or a composite of minorities who have special problems and saying this is the New Politics, because this is as old as the history of politics in this century. This is the kind of politics of the first act, in which he said, 'Here is the problem. We have the problems in the cities. We have problems in the country. We have a balance of payments problem. We have a problem in Vietnam.' But where do we go from there? Well, we use the same old rhetoric, at least in his campaign. No new language—'We're going to do better, we're going to get the country moving, we're going to do more—' but beyond that, no specification as to which way the country's going to move, or what we're going to do more of, or what or who we're going to do better.

"And really he's brought no new politicians to the scene. We've brought back Pierre Salinger from retirement. It's rather amazing where all the Knights of the Round Table went in the interim, and what they did. Ted Sorensen's come back. He's not exactly a new politician. Larry O'Brien, Kenneth O'Donnell, and Arthur Schlesinger is back as an adviser. He's said nothing about some of the old politicians who never left, or at least those officeholders, no indication of any particular criticism of Dean Rusk, and I think there's reason to criticize Rusk; no criticism of the former Secretary of Defense—even at one point a willingness to use a kind of selective endorsement by the Secretary of Defense; no criticism of General Hershey; not even a suggestion that we might look at the operation of the FBI, or its director.

"In fact," McCarthy went on—to the great glee of his followers hearing the quiet man speak thus, "we've had a return to the old politics with his entrance into this campaign. One, the misrepresentation of an opponent's record. Secondly, the device of trying to detract people from a consideration of the issues in the recent suggestion, in advertisements here that the real contest, he says, is between him and the Vice President. I don't think that's the real contest. He's even brought the cocker [sic] spaniel into the campaign. I don't see that this clarifies the issues very much or helps clarify them . . . and also, yesterday, a cold plunge in the Pacific. This is all old politics. Nothing new, nothing surprising. . . .

"But while he does take the cold plunge in the Pacific and brings the dog and does these other things, up to now he has refused to meet me in debate or in joint appearance. I think that if this is the case in a primary when we're reasonably friendly and of the same party, one must say, will he face Richard Nixon in the general election—or, if elected, would he be prepared to face some of the tougher debaters around the world? And more importantly, perhaps, would he be prepared to face the mistakes that he and others have made in the past?"

McCarthy clearly was in contention in this one, and climbing. The next day, when both he and Kennedy toured the Portland zoo, they came within 50 yards or so of meeting, and Kennedy reacted like a man escaping the plague. While he was walking down a hill, shaking hands and talking to a reporter, Bill Barry came up and informed him of McCarthy's approach. "Let's get out of here," Kennedy said; he raced to his open convertible, hopped in and sped away—to cries of "Coward! Chicken!" from banner-waving McCarthyites. Why Kennedy had bolted—whether he simply had been momentarily rattled, or whether he still was determined not to give McCarthy additional exposure in what certainly would have made a page-one picture and story all over Oregon—wasn't clear. But whatever the reason, Kennedy made McCarthy look good. The Minnesotan himself, in his easy, ambling style, strode over and boarded the Kennedy press bus—the candidate's departure having been so quick and unexpected that the bus driver and press had been caught flat-footed. In good humor, he chided the reporters covering Kennedy and invited them over to his campaign.

Earlier, Kennedy had spoken at the Neveh Shalom Synagogue, reiterating his view that "in Israel—unlike so many other places in the world—our commitment is clear and compelling. We are committed to Israel's survival. We are committed to defying any attempt to destroy Israel, whatever the source. And we cannot and must not let that commitment waver." The words attracted little attention. He had said them before, and besides, he and McCarthy agreed on Israel. Shortly before the near-confrontation at the zoo, McCarthy had addressed the annual picnic of another synagogue, and had made the same point. But Kennedy's talk was included on a television report that was seen later around the country. According to an Egyptian correspondent named Mahmoud Abel-Hadi, the sight of Kennedy wearing a yarmulke in the temple and uttering those words greatly upset one young man in the Pasadena, California, home of a Jordanian family. The correspondent reported that Shereif Sirhan, brother of the youth in question, named Sirhan Bishara Sirhan, had told him: "He left the room putting his hands on his ears and almost weeping."

After another shopping center stop and high school rally marked by fairly large but not particularly attentive crowds, Kennedy flew to Eugene for a reception and an early start on the final campaign day. On Monday, May 27, again using two small planes, he hopped across southern Oregon, where an estimated 11 percent of the voters lived, in the hope that his rural magic of Nebraska might take hold. The judgment at this time was that Portland and the Willamette Valley remained crucial, but that there was little more Kennedy could do there.

While McCarthy was triumphantly addressing a large downtown Portland crowd and then walking through a sea of well-wishers back to his hotel, the Kennedy party was shaking off the effects of another ominous start.

Flying into Roseburg for the first stop, the two planes narrowly averted a crash when they approached the runway for landings from different directions. Nobody was hurt, but the incident did nothing for already shaky nerves.

In Roseburg, the heart of hunting country, Kennedy was confronted with signs reading PROTECT YOUR RIGHT TO KEEP AND BEAR ARMS. Grimly, he said, "I see signs about the guns. I'm wondering if any of you would like to come and explain." A heavy man in a lumberjacket, who identified himself as S. J. "Bud" Schoon, a local businessman and a director of the Association to Preserve Our

Right to Keep and Bear Arms, Inc., spoke up, reluctantly. He called pending federal legislation a "backdoor bill for the registration of guns." Kennedy tried to persuade him otherwise. "All this legislation does is keep guns from criminals, and the demented and those too young," he said. "With all the violence and murder and killings we've had in the United States, I think you will agree that we must keep firearms from people who have no business with guns or rifles."

A man in a cowboy hat booed, as children waved the signs around him. "Nazi Germany started with the registration of guns," another called out. Kennedy noted that the registration of automobiles and drug prescriptions had not destroyed democracy; he didn't see why gun registration would. But the crowd was not impressed.

That night, to close the Oregon campaign, Kennedy and his wife, along with Mrs. Green, attended a large reception of the faithful at the Sheraton Motor Hotel in Portland. The crowd seemed confident, but Kennedy was transparently pessimistic. The final Quayle poll, just out, gave him a paper-thin edge: 34 percent to 32 for McCarthy, 10 for Humphrey, 9 for Johnson, 4 for others and a critical 11 percent undecided. In past primaries, McCarthy had won most of the undecideds.

Instead of the customary rousing get-out-the-vote speech, Kennedy spoke more like a man who already had lost, and who was trying to soothe the feelings of his workers and find some consolation for all they had done. "I don't know how it's all going to turn out tomorrow. The polls say it's like that," he said, holding two fingers tight against each other. As if to emphasize his yearning for the days of the old politics, he told about "a great-uncle who had voted 143 times against my grandfather Fitzgerald. I guess I wasn't supposed to tell that." He smiled.

If Kennedy was down, Mrs. Green wasn't. Rather than let the crowd go away in a pessimistic mood, she delivered an old-fashioned rouser. "The question is not who started where," she insisted, "but who's going to end up successfully on November fifth. Bob Kennedy's voice has been a lonely voice as far back as 1965. . . . The question is, who is going to have a realistic chance of being nominated in August? . . . In the next twenty-four hours, we can really make a difference." But Robert Kennedy, standing benignly next to her, didn't seem to think so, and for good reason. His technicians, led by O'Brien, had made it clear to him that he was up against a formidable array of difficulties: McCarthy's own effective organization and identification with the Vietnam issue; the organized labor effort in behalf of Johnson-Humphrey; the old Senate Rackets Committee residue of ill will; the highly motivated, independent-minded electorate turned off by emotionalism in its candidates; the poor start resulting from poor organization; finally, the absence of the Kennedy base. To reporters too, O'Brien—kidded as the master of artful poor-mouthing—passed the same gloomy word. "It's a lot easier to poor-mouth when you don't have to," he quipped to skeptical newsmen.

Again as in Nebraska, the demands of the next primary did not allow the luxury of waiting around for the Oregon returns. Kennedy flew to Los Angeles late Monday night and spent Oregon primary day touring the Santa Barbara-Ventura area. The voting to the north, of course, was on everybody's mind, but Kennedy plugged away. Late in the day, a CBS sampling of 400 key precincts flashed to him indicated he was losing. Just before his plane took off for Oregon that night, shortly before 8:30 o'clock, Dick Dougherty of the Los Angeles *Times* brought in the word. Dougherty and Dutton both had been calling for the network projections and Dougherty got back to the plane first. As he started up the stairway, he met Kennedy coming down. "Bad news, Bob," Dougherty told him. "Oh," said Kennedy. "That's too bad." And he turned and went back into the plane.

On the return flight to Oregon another reporter approached him. "I've lost," the candidate told him. "I'm not one of those who think that coming in second or third is winning." No Kennedy ever thought that, or had lost in twenty-six straight elections. But the first defeated Kennedy spoke without rancor, sitting with his collar open, tie pulled down, shirtsleeves rolled up, sipping a beer and eating steak. Humphrey, not McCarthy, was the big gainer, he said. "I think what he [McCarthy] wanted most was to knock me off. I guess he may hate me that much."

Kennedy said he was concerned mostly that he may have miscalculated the mood of the public; he thought they were ready and willing to take steps to deal effectively with poverty and to bring about racial reconciliation. The Oregon defeat, he conceded, would make it extremely difficult to deal with the big nonprimary states and their Democratic leaders. "They will use Oregon as an excuse for not supporting me," he said. Then the candidate, the first time in his life a losing candidate, walked through the plane, thanking aides, trying to cheer up Ethel and others, while Dutton wrote out a congratulatory telegram to McCarthy. Kennedy had insisted on that; that McCarthy never had sent him one always had rankled him.

In Portland, at the airport and in the lobby and ballroom of the Benson, they came in droves to see how a Kennedy took defeat. Some cheered for him, others just looked, almost all pushed toward him. Ethel, wading through the crowd with her customary disregard for her own condition, wore a brave half-smile. Even now the full dimensions of the setback were not known. It looked as though Kennedy would get 40 percent of the total, but he fell under that in the final figures: McCarthy 44.7 percent, Kennedy 38.8 percent, Johnson 12.4 percent, Humphrey 4 percent, others 0.1 percent. As predicted, the lack of the have-not base was irreparable; McCarthy's pull among the suburban electorate was solid and dependable. Kennedy actually won more convention delegates than McCarthy, but the popular vote was what counted in the major effort—to demonstrate voter appeal.

In the lobby also were many supporters of Richard Nixon, who had just scored another sweeping victory over shadow candidates Ronald Reagan and Nelson Rockefeller—and his own past—in the Republican primary. He had addressed a cheering crowd of backers certain that they were witnessing a key juncture in a Republican year of comeback.

Kennedy took an elevator to his suite, then came down to the Mayfair ballroom on the mezzanine floor to thank his supporters, some of them in tears, and to read the telegram to McCarthy. "My sincerest congratulations on your victory here in Oregon," it said. "We can both take some satisfaction in the overwhelming expression of the Oregon voters for a change." In light of the defeat, Kennedy told the glum crowd, he was taking a look at his organization, and "I have decided to send Freckles home." "Don't do it," a man shouted. "Not really," Kennedy said, smiling. In a clear and unwavering voice, he thanked the voters of Oregon for making "a fair judgment, as they were supposed to do," and the newspapers, radio and television for their fair coverage. He repeated his admiration for the "majestic beauty" of the state, expressed the hope he could return again, and that out of the discussion of issues in the Oregon primary the United States "will move closer to peace at home and around the globe." And then he left, as he had come in, through the hotel kitchen. But because he had lost, there was no press conference.

Over at an old Elks Club building, McCarthy appeared before a rejoicing mob. A cheerleader led the victory cry: "Gene in '68! The rest will have to wait!" The McCarthy dropouts and the "teenies" went around hugging and kissing everybody who had had any part in the great adventure.

"In Nebraska," he said, grinning as he had not done on election night since New Hampshire and Wisconsin, "we discovered our weaknesses. In Nebraska we discovered the weaknesses of the opposition and we were ready for the western movement. . . . Every wagon train gets as far as the Missouri River, but the real test starts up the Oregon trail." McCarthy didn't eleborate on what his and his opponent's weaknesses were, but the audience didn't need clarifications. "We know who had the best horses and the best wagons and the best men and women, and I think we proved that in Oregon." Cheers and hundreds of arms waving the V peace sign. "The next test is the California trail," he continued, "and we're on to California. California here we come!"

When the whistling and applause died down, McCarthy got in his licks at the press. "I haven't had any reporter ask me who I'm going to yield to in Chicago. . . . I don't think we'll be asked that question again in 1968. I think we've answered it in Oregon. . . . It was just a question of finding our constituency. . . . Our campaign here didn't bridge the generation gap. It was solid all the way, and it will be solid all the way to Chicago. . . . The President said we will have riots in the streets this summer. Instead of riots, we'll have singing and dancing in the streets. . . . We'll take the fence down around the White House and have a picnic on the lawn." The crowd burst into "The Battle Hymn of the Republic"—triumphant, ecstatic, as the V signs waved in rhythm.

Over in Kennedy's suite on the seventh floor of the Benson, the usual crowd of aides, friends and newsmen who always gathered for a victory party drifted in, went to the bar for a drink and milled around in quiet conversation. There was a long living room, with two bedrooms off to the right. The candidate was in his shirtsleeves, tie off but still wearing the PT-109 tieclasp that had been the symbol of Kennedy political invincibility. He held a heavily watered-down drink in his hand, and he was in a quiet, reflective mood—not bitter, not shocked, not even transparently disappointed, just resigned. Others in the room moved deferentially toward him, spoke a few words of encouragement or tried to say something light, then moved on, not wishing to intrude excessively on this awkward time. One reporter joked that there was a movement to eliminate the word "viable" from the English language. "No," Kennedy said, shrugging his shoulders, "it's true." He blamed the defeat on himself, on his own deficiencies as a campaigner and an organizer.

At one point he moved into one of the bedrooms, where aides like Sorensen, Goodwin, Salinger and Dutton, plus the newer team members who had not tasted the sweetest Kennedy victories of earlier years, held a postmortem. The discussion was, of course, California, and what had to be done to make up lost ground. It was said, or assumed by all, that Kennedy now would have to debate McCarthy. The old arguments about giving exposure to the opposition and the traditional no-debate stance of the front-runner didn't apply anymore, and nobody had to make them. McCarthy's forces in California were capitalizing on their "Why won't Bobby debate" slogan, and it had to be countered. Also, McCarthy in every primary had run stronger than the final polls indicated he would. This meant he was pulling in most of the late undecideds. A good showing in a debate could check that pattern. The decision would be announced the next morning, on arrival in Los Angeles.

Kennedy came out again, quietly cordial, fielding questions with patience. Ethel, as unassuming as ever, talked with friends and acted for all the world that losing was routine for a Kennedy. The defeated candidate, now sitting on the arm of a chair, said he had known he was in trouble when he had failed on the visit to the Omark plant on May 17 to draw a responsive reaction from the workers. He could tell, he said, they weren't tuned in, that he wasn't communicating with them. He asked aides where he had run well and where poorly, and took the information as much like the campaign manager he once was, matter-of-factly, as like the disappointed candidate he now was. When a reporter asked him if he thought the Oregon defeat had hurt him, the stupidity of it made him laugh. "It certainly wasn't one of the more helpful developments of the day," he said. It went on like that, just conversation with everybody talking around the edges among themselves and with him, and giving him a wide berth.

Three reporters, however, drifted over and began questioning him about his plans and about this and that detail, until it began to take on the semblance of a press conference. After taking the questions for a while, Kennedy finally looked up and said, "We'll be having a press conference tomorrow. Can we hold off until then?" The newsmen agreed, and two walked away. I stayed and asked him: "Why do you put up with all these questions at a difficult time like this?" Kennedy reached over and put his arm around me. "Because," he said, "I like you." And he walked away. It was one of those moments newsmen had to guard against in order to do their job. Kennedy had a way of pulling individuals around him into his orbit, a strange disarming quality about him that somehow evoked sympathy for the man who had everything. It was not difficult, exposed to this quality, for those around him to believe that to deny him, at forty-two, the leadership of the most powerful nation in the world would be unjust. It was their weakness, and an element of his power.

The candidate obviously was weary, and so Dick Tuck began clearing the room. He reminded everybody of the early departure for California, and Robert Kennedy went to bed, a loser for the first time. Oregon had been a blow, more than most realized. Not only had he lost a primary and seriously diminished if not destroyed his basic argument to the power brokers, that he was the people's choice; he knew now he could not win the kids—the young activists his own indecision in late 1967 had sent looking elsewhere, and McCarthy's victory in New Hampshire had captured. With them behind him in Oregon, and with the war issue to himself, he might have made up for the absence of the Negro-blue collar base. Robert Kennedy approached the California primary knowing that this base would be back working for him. But he knew too that many of the kids for whom he had both a political and emotional need were gone for good now. Not only gone, but committed more enthusiastically and vigorously than ever against him.

CHAPTER SEVEN

California

THE COMEBACK

THE IDEA OF a Kennedy losing was hard to grasp, after all the victories and all the adoring crowds, and for no one was it more difficult than for Ethel Kennedy. As the campaign plane flew south to Los Angeles on the morning after the Oregon defeat, she walked up and down the aisle, trying to be lighthearted and philosophical about it, yet transparently rationalizing the setback away. "Such a small state," she told several reporters "and only 15,000 votes [the final margin of defeat was nearly 20,000]. Do you really think it will make that much difference, if we win in California? That's the important state, more like the rest of the country."

But her husband and his stategists knew full well they had been gravely damaged in Oregon. Just winning in California probably would not be enough. Yet winning big—that is, a clear majority of 50 percent or more—did not seem in the cards either. Once again, it was a three-cornered race, with Kennedy and McCarthy splitting the anti-Administration, antiwar vote. This time, neither the President's name nor Humphrey's was on the ballot, but there was a pro-Administration slate running under the banner of State Attorney General Thomas C. Lynch, California's highest Democratic officeholder. Before Johnson's March 31 withdrawal, Lynch had been persuaded to serve as a stand-in for him, and many regular Democrats who would have preferred Kennedy agreed to run on the Lynch slate. Now, it provided a vehicle for intensive labor activity against Kennedy.

In her description of California as more important and more typical of the country than Oregon, Ethel Kennedy was correct. Although the Oregon defeat was a crushing blow, an impressive Kennedy victory in California, enabling him to pick up 172 of the state's 174 delegate votes, would not be dismissed out of hand by the party brokers in the other major states. And a powerful demonstration of Kennedy pulling power in California's cities, rooted in his have-not coalition, would all but assure the party a similar base in key northern industrial areas in November. It had become a political cliché that California was a microcosm of the country, with its industrial, agricultural, racial, ethnic, and cultural mix; yet the idea that there could be anything microcosmic about this vast and sprawling state of pell-mell growth was a contradiction in terms. California was a huge slice out of the national pie, 20 million residents or one-tenth of the nation, growing twice as fast as the country as a whole; 7 million registered voters, 4.3 million of them Democrats. He who established political strength there in a test at the polls brought a persuasive argument to the party convention. Coupled with his own New York, largest of the delegations with 190 votes, Kennedy would have 28 percent of the 1,312 votes needed for the nomination from these two states alone.

In terms of mounting a campaign, California is more like a nation than a state. It stretches 800 miles north to south, with three great population centers along the west coast—San Francisco-Oakland in the north-central portion, Los Angeles-Long Beach in the south and San Diego at the southern tip. From each, a huge suburbia spreads east into the valleys of the center. Los Angeles County, with a Democratic registration of nearly 1.5 million Democrats, 38 percent of the state total, is the key; its concentration of lower-income minority groups, particularly Mexican-Americans, makes it a magnet for street campaigning by a charismatic candidate. It is possible for a motorcade to roll all day through city streets without seeing the open road. And with jet travel, a campaigner may sweep from one great population center to another in a day, and back again.

Yet the fantastic suburban growth requires also that a candidate reach out from the major centers; down through central California, from Sacramento to Stockton to Modesto to Fresno and Bakersfield. Beyond the three great urban centers, and excluding the Sierra Nevada range and the great deserts beyond and south, it has been said California is one gigantic suburb of the good life.

Some 86 percent of the state's population is considered urban, but the suburban sprawl offers a distinct political challenge from that posed in the city centers. In California, life in the suburbs is not generally as placid or insulated as it is in Oregon. The economic level varies greatly and so do the ethnic and racial strains; one suburb may be merely a geographical extension of the city mix, with more breathing room; another may be an enclave of affluence, snobbery and button-down red-neckism or, in one of California's many college and university towns, an oasis of quiet social and intellectual nonconformity. For the politician, each is a different challenge and a different opportunity, and must be reached.

Even with the speed of modern air transportation, however, no candidate can blanket California in a few weeks. For that reason, it is known among the professionals as a "media state"—one in which television, radio and newspaper advertising and free exposure are much more important than in those states whose population is small and concentrated. This fact imposes great financial, logistical, specialized manpower and scheduling problems on a campaign, far beyond what exists in states like Nebraska, Oregon and Indiana. Money must be found for television time and radio and newspaper space; writers for television, radio and newspaper ads must be corraled. Itineraries must be worked out that not only put the candidate in contact with the maximum number of likely voters; they also must provide lively, colorful and different activities that will induce cameramen to shoot footage and will persuade television directors and producers to use the film on their mass-audience news shows. Above all, the candidate must be visible. Where there are great blocs of voters, it helps if he is seen in the flesh; where he cannot go personally, he must be brought to that great unseen force in a major campaign—the millions who sit in darkened parlors sipping beer while the flamboyant motorcade roars down other streets to the cheers of the more demonstrative, but less numerous, citizenry.

For Robert Kennedy, needing an immediate comeback to counter the blow sustained in Oregon, California offered what seemed an ideal opportunity. There was no other place in the nation where so many delegate votes were riding on the outcome of a single primary; there was none where the enthusiasm for him in earlier visits had been as high; there was, as in Indiana and to a lesser extent in Nebraska, the Kennedy base of the disadvantaged and/or disenchanted. This time he had not only the blacks of Watts and San Francisco's Fillmore District; in Los Angeles, in San Diego, in the Central Valley he had the Mexican-Americans, who already had demonstrated their affinity to him. Kennedy's association with Cesar Chavez, the heroic organizer of grape workers, gave him a powerful and diligent ally in this group, which constitutes 10 percent of the population, larger than the state's Negro population of 7 percent. And on top of that, more than one of every three Californians is under the age of eighteen, and nearly one in every two is between eighteen and fifty-five. None in the first group could vote, of course, nor could all in the second, but in terms of their impact on California's thinking, they were influential. There seems to be in the state a near-worship of the young, of being or wanting to be young, and of acting accordingly; such an attitude was likely to redound to the advantage of the forty-two-year-old Senator who looked even younger.

For all this, however, two other elements in the state's demography posed potential difficulties for Kennedy, difficulties that were compounded by McCarthy's victory in Oregon. One was the suburban sprawl, and the independence in politics it often bred. In Oregon and in earlier primary states, McCarthy had fared exceedingly well in the suburbs and would be a threat in California for that reason. Also, large pockets of old liberals of the Adlai Stevenson stripe and of new radicals whose anti-establishment activism fired up Berkeley and other campuses, were pulled to McCarthy. On the war, the burning issue, he had been with them first, and he had brought the hated Johnson down; along the way to California he had eschewed the old politics, or so it seemed to them, while Kennedy unabashedly tried to work both sides of the street.

The fight for California, of course, had been going on long before the Oregon defeat. Jesse Unruh's early confidence that Kennedy could win this key state with or without Johnson in the primary had been a factor in the decision to run. Kennedy's first full weekend swing after his March 16 entry into the race had been an all-out, wildly successful jet tour through the state. In all, he already had been in California as a declared candidate for all or part of ten days when he returned on Wednesday, May 29, to make his final week's push for political survival. Unruh had been installed as the leader of the home-grown forces, and the top Kennedy men available had been shuttled in and out all spring. Steve Smith, the campaign manager, spent much time in the state;

Frank Mankiewicz, the press secretary, turned that job over to Dick Drayne and Hugh McDonald in Indiana and headed for California, his home state, to become an operations man and tactician; other "Easterners" in numbers that sometimes irritated the Californians moved into the state in the classic Kennedy style.

Because the campaign had been put together literally on the run, and because there were differences of opinion and personality between the Californians and the superimposed Easterners, the Kennedy operation was not the "juggernaut" it was cracked up to be. This was true in the earlier primaries too, but at least in Indiana and to a large extent in Nebraska, the out-of-staters were able to come in, take a reading of the situation, and go to work. In California, there were knowledgeable workers on the scene who already had taken their own reading, and the complexity and size of the state resisted quick analysis and solutions by the Easterners. Frictions resulted that compounded the candidate's task as he approached the crucial stage.

The problem was largely solved when Steve Smith in the last three or four weeks planted himself in the Los Angeles headquarters, with an office next to Unruh's. (Steve Smith's arrival necessitated reference to him as "Steve Smith East" to differentiate him from an Unruh aide with the same name, who from then on was called "Steve Smith West.") A dual clearance was instituted for most major matters, and although it was time consuming and caused some bogging down, some sensitivities were eased. And in the end it was Smith by his firm hand who guided the campaign. In the north, John Seigenthaler ran an effective operation, reportedly with less difficulties with the locals.

Organization problems aside, though, there always was confidence about California in the Kennedy camp. In Larry O'Brien's eyes, the essential thing was there—the Kennedy base. It was there, it was highly motivated, and the job was to get it to the polls on primary day. Had Kennedy squeaked by in Oregon, there might have been relatively little to worry about. But he hadn't, and as a result McCarthy was coming on; his appeal to the middle class could not be dismissed in this "one gigantic suburb." It had to be countered.

The obvious way was to hit the have-not areas hard—Watts, Fillmore, East Los Angeles, other strongholds of Negro, Mexican-American and other ethnic populations. It had worked in Indiana and Nebraska, and there was no reason why it wouldn't work in California. If to this base could be added the old liberals, that is, if they could be weaned away from McCarthy, Kennedy would win—perhaps handily. Some of the Easterners felt it could be done. Mankiewicz organized an impressive roster of volunteer speakers who worked this lode, with some success—Arthur Schlesinger, Ted Sorensen, Michael Harrington, Abba Schwartz, Marietta Tree, Daniel P. Moynihan, Harry Golden, Alexander Bickel, Adam Yarmolinsky, others. They met with special voter groups, and did radio and television shows in both northern and southern California.

But many of the Californians believed that the old liberals were locked in on McCarthy—as they proved to have been in Oregon—and that it would be better to look elsewhere for strength. The elsewhere they had in mind was the suburbs themselves, and they contended that the way to counter McCarthy's strength there was to challenge it head-on. Some Easterners, including O'Brien, agreed. The only trouble with this mix was a concern that the pursuit of the have-not city vote would hurt Kennedy in the well-off suburbs. Excessively enthusiastic crowds, with many black and brown faces in them, scared the entrenched whites. So did the frantic quality of the motorcades, with the candidate being clawed at, screamed at, pulled from the top of his car. Still, in every state except Oregon, Kennedy did best in those areas where he campaigned personally. In this final week, he and his top advisers decided, he would campaign hard among both the have-nots and the suburbanites, while attempting through television in this "media state" to project an image of quiet competence. Maybe now that he was a loser, he would be free of the "ruthless" albatross at last, or at least have a chance to shake it.

The first step toward a comeback, and to stem McCarthy's rush, was the debate. It could help combat the ruthless image and cut into the late undecideds. As the Kennedy plane eased into Los Angeles International Airport, the mimeograph machines in the rear were grinding out a statement expressing his willingness to take on the Minnesotan. Kennedy released it at a press conference in a small auxiliary terminal building. Acknowledging that the Oregon loss represented "a setback . . . which I could ill afford," Kennedy said the results there nevertheless showed "that the citizens who make up the Democratic Party wish to select their own nominee—since they gave more than eighty percent of their vote to candidates willing to enter the primaries and anxious for a change." If Humphrey were to be nominated to run against Richard Nixon, Kennedy said, "there will be no candidate who has opposed the course of escalation of the war in Vietnam . . . [and] no candidate committed to the kinds of programs which can remedy the conditions which are transforming our cities into armed camps." In a roundabout way, he was saying that unless he or McCarthy stepped aside after California, Humphrey was sure to be nominated and the status quo continued.

"It is hard to believe after the recent months of hope," Kennedy said, "that our political system will ultimately fail to offer the people a chance to move in a new and more hopeful direction. I am sure Senator McCarthy believes his qualifications for the Presidency are superior to mine, as I believe myself qualified for the office. And I will be glad to discuss these matters, separately or in joint appearances."

California, Kennedy noted, "contains all the diverse elements of the national electorate" and was the largest state in the Union. Slates committed or leaning to all of the three major candidates were running. "It will be a fair test of each candidate's appeal [and of] our differing philosophies, convictions and qualifications . . . and I will abide by the results of that test."

This last could be taken only one way: if Kennedy lost in California, he would step aside. Asked whether this was a fair interpretation, he said, "I think the statement speaks for itself." Did he mean California was the ultimate test of his grass-roots candidacy? "Yes, that comes close to the way I would describe it." Why debate in California when he wouldn't debate in Oregon? "Because conditions have changed . . . I'm not the same candidate I was before Oregon and I can't claim that I am." Well, what about the "viable" statement? "I slept on it." If he lost in California, would he take the Vice Presidential spot? No, he would go back to the United States Senate and busy himself "raising the next generation of Kennedys. It's a full-time job." How did it feel to be the first Kennedy to lose? "I feel like the man Abraham Lincoln described who was run out of town on a steel rail, and said, 'If it were not for the honor of the thing I'd rather have walked.' " And then, to end the press conference: "I've got to go because I have a lot of fans waiting—I hope."

In downtown Los Angeles, Kennedy's fans—and the dependable advance work of Jerry Bruno and Jim Tolan—assured that he would not be disappointed.

The motorcade sped to the heart of a Mexican-American, Negro and business district, and the turnout was colossal, even by Kennedy standards. Ever since the candidacies of Adlai Stevenson, this area at the noon hour had been sure-fire for huge Democratic crowds, and the narrow street helped convey the impression of great numbers. Knowing these facts was part of being an effective advance man and Bruno knew the best crowd locations like the professional fisherman knows where to drop his nets. Ticker tape and torn newspapers poured out of the office windows high above him as Kennedy made his first venture among the voters as a loser. Rafer Johnson, the Olympic decathlon champ from Los Angeles, was at his side now, and again the hands reached up and out—black and brown and white hands, young and old, and faces upturned, some of them calling encouragement, others seeming to try to convey it with their looks of adoration, or of simple gladness. A woman in her fifties, one of those itinerant crackpots who show up at rallies all over the country, ran alongside the car dressed in an outlandish costume of green sequins and long, dangling beads, shouting "Pooh on Oregon!" and administering a rousing Bronx cheer. (To those in the press bus who had been traveling too long, she looked suspiciously like "Gladys," but even on her worst mornings after, "Gladys" was in better shape.)

Kennedy, fist punching against hand, was back to his early campaign days of all-out fervor. "I need your help!" he barked. "I've come to ask for your assistance!" And the cries came back: "You've got it!" "Sock it to 'em, Bobby!" The crowd grew so thick that many times the motorcade was obliged to stop to avoid hitting spectators. Police cars, parked at intersections, reported to their stations that all the rules were being broken, the motorcade was going through red lights, and maintaining order was hopeless. "We're down to Eighth and L," one patrol car driver reported. "There is complete disregard."

According to the Kennedy staff, police escorts were denied by the City of Los Angeles because regulations did not cover candidates for party Presidential nominations. Los Angeles officials contended that the Kennedy party refused the escort. At any rate, the patrol cars finally began to cruise behind Kennedy's car, shouting over their loudspeakers: "Move out of the street! Don't get hurt! Please stay on the sidewalk before somebody gets pushed under a car and crushed." But none of it did any good. Hundreds now were walking alongside

Kennedy's car, trying to climb onto it or the photographers' convertible just behind. After half a block or so Kennedy would stop and speak—his short, intense plea for support. He had his coat off now in the hot midday sun, and was drenched in perspiration. He never wore an undershirt but he changed shirts two or three times a day if he could. "Will you give me a hand on June 4?" he asked, and more hands reached out to him. His cufflinks went, his tieclasp, and his shirttail was pulled out. He beamed with satisfaction and hope. "These are my people," he often would say to Fred Dutton as the motorcade approached a group of smiling Negroes; he had good reason to say it on this emotional day.

As the motorcade worked its way out of the downtown area to the freeway, Kennedy stopped his car, got out and ran back to the photographers' car, grinning like a small boy. "From now on," he told his camera-toting friends, "Los Angeles is my Resurrection City!" And he raced back to the lead car. At a reception for workers at the Beverly Hilton Hotel, the candidate repeated the line, but in words that later would seem even more poignant: "If I died in Oregon, I hope Los Angeles is Resurrection City." Then, running far behind schedule, the caravan moved out into heavily populated El Monte, Fontana, San Bernardino, Riverside. Kennedy returned around midnight, weary but elated. Oregon may have been a political aberration, nothing more, and could be placed in proper perspective in California.

McCarthy, who had arrived in the San Francisco area aboard a plane bedecked with colored balloons and a champagne-sipping staff and press corps, spent the day in triumph. "There's a line of poetry about summer patriots," he said in Stockton. "There should be something about spring and summer candidates. . . . They should have been with me in New Hampshire." In the euphoria of the moment, Gerald Hill, head of the California Democratic Council and one of McCarthy's California managers, pronounced Kennedy "eliminated" from the race. As for the debate, which McCarthy had long sought, the Minnesotan seemed still willing but less than hopeful about what it would achieve.

On Memorial Day, Kennedy whistle-stopped through the Central Valley—Fresno, Madera, Merced, Turlock, Modesto, Stockton, Lodi, Sacramento—in a bid to cut into the suburban vote. As usual he directed his fire at Humphey's "politics of joy," but now that he was going to debate McCarthy, he no longer ignored him. "I was the first one of any of the present candidates who ever spoke out against the course we have been following in Vietnam," he told trainside crowds. It was true that Kennedy had been prominently on the record against Vietnam earlier than McCarthy; inexplicably, he never bothered to say so in Oregon. Now all the chips were on the board, and any reluctance he had to treat McCarthy as his serious foe had been wiped away by Oregon.

McCarthy, for his part, continued the politics of disdain. In defeat or in victory, it was the same; he could not be accused of inconsistency. Of Kennedy's hint that he might quit the race if McCarthy beat him in California, the Minnesotan told his cheering backers at his southern California headquarters in Los Angeles: "Really, I don't have the right to withdraw. . . . Apparently there are some candidates who come into this campaign on their own and can go out of it." But for his part he had entered the race when others asked him, and "no matter what happens here we'll carry the issues to Chicago." He likened Kennedy to a small boy who threatens to hold his breath if denied what he wants. The troops cheered and, as he drove off, somebody waved a sign that said GENE THE GIANT KILLER.

The emphasis in the McCarthy campaign in these last days was on radio and television—as much as he could get. His California managers admitted McCarthy still had a recognition problem, and although it had been reduced by his Oregon victory, Kennedy still was far better known. For that reason, the McCarthyites were elated at the prospect of a debate. They knew it would give their man statewide exposure, and they were confident he would make mincemeat of Kennedy in the discussion. The candidate, however, seemed cool to the whole business now that it was at hand. In a radio interview on station KLAC, he complained about the format, in which three network reporters would be asking questions. He had wanted a straight-out debate, McCarthy said, but Kennedy wouldn't do it. "We'd rather have had a real debate confrontation," he told his interviewer. "He [Kennedy] just wants to sit around the table and be nice to each other."

While Kennedy was trying to make inroads in the Central Valley, McCarthy ventured into Watts, where his workers were staging a barbecue in a local park. The residents turned out, but largely ignored him as he spoke. "I'm eating his ribs, but I'm voting for Kennedy," one man said, and many around him laughed and nodded. Although McCarthy had continued to say he would not pitch for any voting bloc, he noted here that he had called the unseating of Representative Adam Clayton Powell "unconstitutional," and he equated "black power" as a legitimate social and political force with "Irish power" and "Italian power."

"There's never been any group in America who had any more reason to organize themselves to get their rights [than the blacks]," he told the 100 or so who were listening. "The question becomes one of how you organize and what you do, but there's nothing contrary [in this] to the traditions of this country." And then McCarthy went off to a Jewish neighborhood where he spoke about civil liberties.

After his whistle-stop tour and a reception in Oakland Thursday night, Kennedy had an unscheduled, unannounced midnight meeting with about 100 members of the area's "Black Caucus," including the Black Panthers, at the Taylor Memorial Methodist Church in West Oakland. The meeting, arranged by Willie Brown, a Negro Assemblyman from San Francisco, was similar to earlier secret sessions Kennedy had held with black militants in Indianapolis and Omaha. Riding to the meeting with John Glenn, Barry, Dutton, Rafer Johnson and Stan Tretick of *Look,* Kennedy turned to Glenn. "These meetings aren't very attractive," he told him. "They need to tell people off. They need to tell me off." He was right, and at the meeting the black militants filled their need. They called him "a white big shot" and worse, to a point where Johnson, angry, tried to intercede. "No," Kennedy said. "This is between them and me." He was aware that such meetings were a catharsis for the black militants, and from them he drew a clear awareness of the intensity of feeling in the black community, and a better understanding of it.

In San Francisco on Friday, May 31, the day before the debate, Kennedy telegraphed one of his punches, revealing some of the same kind of open bitterness toward McCarthy that his opponent had shown in Oregon. McCarthy had griped about ads placed in California college papers "distorting" his Senate voting record; now Kennedy charged that McCarthy "distorts the truth and demeans politics" in a campaign advertisement claiming that Kennedy somehow was involved in the Johnson Administration's 1965 intervention in the Dominican Republic, which occurred after Kennedy had quit the Cabinet. The ad in question said:

"Kennedy was part of the original commitment [in Vietnam]. He participated in the decisions that led us to intervene in the affairs of the Dominican Republic. He was directly involved in the disaster of the Bay of Pigs. And as a member of the National Security Council, he must bear part of the responsibility for our original—and fundamentally erroneous—decision to interfere in Vietnam."

The advertisement, Kennedy protested during a speech on Vietnam to the Commonwealth Club, "completely distorts the facts in connection with the situation in Vietnam . . . [and] the facts in connection with the Dominican Republic. This great crusade that began for the future of the United States is now involved in a campaign to distort me and stop me. . . ."

The bitterness, now on both sides, came at a time when supporters of Kennedy and McCarthy in other states were beginning to work out de facto alliances in struggles to prevent Humphrey from locking up their delegations. In Michigan, Colorado, Missouri and Pennsylvania, joint stop-Humphrey efforts were going forward or had been tried. But as the personal alienation intensified between the two men and between their top echelons, all hope of accommodation was vanishing. On the eve of the debate, which was to be televised nationwide, that hope barely flickered.

McCarthy, predictably, treated the debate off-handedly and declined to cram. He spent the night before the confrontation at Dodgers Stadium in Los Angeles, relaxing from the campaign by watching Don Drysdale pitch a record-tying fifth straight shutout, over the San Francisco Giants—and taking a holiday from the new politics by posing in a baseball cap. Kennedy, meanwhile, had his brain trust—Sorensen, Edelman, Gwirtzman and others— working overtime, compiling facts on the McCarthy record and feeding him background on major issues that probably would come up the next night. He rose late on Saturday and subjected himself to a two-hour skull session, with Sorensen, Schlesinger, Mankiewicz and a few others throwing questions at him and proposing answers. Schlesinger, with a talent for the compact reply, was particularly helpful; Sorensen kept the exercise from rambling; Steve Smith and Dutton screened out the many who wanted to participate. Walinsky flew up from Los Angeles and Pat Moynihan dropped by. Neither was admitted, not because they did not have something worthwhile to contribute, but because time was short and the exercise had to be kept within workable bounds.

On Saturday, while Kennedy continued his cramming at the Fairmont Hotel, McCarthy kept to his schedule of campaigning up the coast from Los Angeles, arriving in San Francisco in the afternoon. By the time he arrived, Kennedy's aides had decided to give him some exposure too, and the appearance of casualness. He went to Fisherman's Wharf for lunch and a round of leisurely handshaking, duly photographed by the television crews. Then he returned to the Fairmont for two more hours in the study hall, and a ninety-minute nap. During the afternoon session, while discussing ghetto problems, somebody off-handedly mentioned what the impact would be of moving Negroes into the strong conservative suburbs of Orange County. Few paid much attention to the throwaway line, but one of those who did, it turned out that night, was Robert Kennedy.

It was a sunny, pleasant Saturday, and other Californians occupied themselves in other ways. Many went to the southern beaches, where young McCarthyites accosted them with blue and white paper daisies to paste on their swimming suits; some played golf and tennis, some went boating, some went shopping for food and other things they would need for activities planned during the coming week. In this latter category was a short, olive-skinned young man with black bushy hair who, accompanied by two friends, went to the Lock, Stock and Barrel Shop at 8972 East Huntington Drive in San Gabriel, part of the Los Angeles sprawl, looking for some armor-piercing .357 magnum ammunition, such as is used by California highway patrolmen. The owners, Ben and Dona Herrick, told the customer they didn't have it in stock, because they didn't want to sell such powerful stuff to sports shooters. Instead a sale of four boxes of .22 caliber bullets was made for $3.99.

That night, the one-hour debate was held at Station KGO-TV, an ABC affiliate, at 6:30 o'clock, Pacific time, and shown live in the East, where it was then 9:30. An hour later, the tape was shown in the Midwest, two hours later in the Mountain states and finally, 9:30 P.M., Pacific time, on the West Coast. Some 171 television and 825 radio outlets presented the show. Crowds, predominantly McCarthy supporters, gathered outside the studio more than an hour before air time. McCarthy, calm and benign as always, arrived first, at about 6:10, to shouts of "We want Gene!" Prior to leaving for the studio, an aide reported, he had been "way up. . . . He's been sitting around for an hour making funnies and singing Irish songs." Just inside the lobby, McCarthy was confronted by a small group of Negroes from a local newspaper demanding representation on the panel, which was composed entirely of white ABC reporters—Bill Lawrence, Frank Reynolds and Bob Clark. "Television can make a contribution to solving the racial crisis," one of the protestors said to him. "I know it can," McCarthy said, and continued walking inside.

About six minutes later, Kennedy arrived in a closed car with wife Ethel, Bill Barry, Dick Goodwin, Dick Tuck and Jim Tolan. Before leaving his hotel, an aide said, Kennedy had been lounging in a bathrobe, listening to staff assistants. It was quiet at first, and then the McCarthy chant started again. A crush occurred at the narrow doorway, and Ethel, trying to force a smile, finally squeezed through. The protestors confronted Kennedy. "You appeal to Negro people," one said, urging him to intercede. Kennedy responded: "Is there a Mexican-American on the panel? Is there an Irishman?" The man was taken aback. "It's obvious all you want is the votes, Senator," he said. Kennedy, making no reply, moved on.

The scene of the debate was a standard television studio, with a semicircular backdrop and three cameras focused on five seats around a small table. Although there were more than 100 seats in the studio, only a small pool of reporters was admitted. The rest watched the show on monitors two floors above, where separate viewing rooms were set up by the McCarthy and Kennedy traveling parties for "their" reporters, an indication of the kind of liaison that existed by this time between the two camps.

With Reynolds as the moderator, the panel members alternated in asking questions of the candidates. Each had the right of reply to the other's first answer. The subjects were predictable: Vietnam, urban unrest, the performance of the Johnson Administration.

At first blush, it seemed to be rather a polite fencing match, with neither candidate having much edge. The answers of both men were familiar; so few sparks were being generated that at one point Reynolds remarked, "There don't seem to be too many differences between Senator McCarthy and Senator Kennedy on anything, really."

But while McCarthy remained his benign self, seemingly trying to avoid sharp differences, Kennedy was following a clearly defined battle plan, worked out in hours of skull sessions and designed to accomplish three things: establish his experience against McCarthy's ten-year age advantage, demonstrate his grasp of the issues and dispel the ruthless image. Kennedy put McCarthy on the defensive almost at once. After McCarthy had opened by calling for "a coalition or a fusion or a new government of some kind" in South Vietnam, Kennedy said: "I would be opposed to what I understand Senator McCarthy's position would be of forcing a coalition government on the government of Saigon in coalition with the Communists, even before we begin the negotiations." To the uninformed, it could have sounded as though McCarthy wanted to hand South Vietnam over and Kennedy did not. "I didn't say I was going to force a coalition government on South Vietnam," McCarthy replied. "I said we should make clear that we are willing to accept that." But he was placed on the defensive by Kennedy's remark.

Next, newsman Clark raised the question of the McCarthy advertisements about Kennedy's involvement in Vietnam and the Dominican Republic. Kennedy seized the opportunity. He drew an admission from McCarthy that aspects of the advertisement were erroneous and that it had been withdrawn on McCarthy's order.

But when McCarthy similarly tried to put Kennedy on the defensive about his own television advertising, Kennedy blandly brushed the effort aside. Lavish praise of Kennedy by former Secretary of Defense Robert S. McNamara had been recorded by the Kennedy staff for use even as Kennedy was blasting the Vietnam policy McNamara had implemented. McCarthy repeatedly had sought to nail Kennedy on this inconsistency, and now he tried again with an oblique reference to the McNamara tapes. "I don't know to what he is referring," Kennedy said unabashedly. Groans were heard in both press viewing rooms, but it is doubtful the vast television audience knew the background, or cared.

At every turn, Kennedy poured in references to his experience. Example:

"While I was a member of the National Security Council for three and a half years, I was involved in some of those matters during that period of time in connection with Latin America and in connection with Africa, with the Test Ban Treaty, and I suppose the most critical crisis that has ever been faced by mankind, the Cuban missile crisis in October of 1962." And: "As Attorney General, I was the chief law enforcement officer and I know how important it is that we abide by the law . . . when I was Attorney General of the United States, we had a riot at the University of Mississippi, and yet no one fired a gun, the marshals or the troops that came into that area. We kept a mob under control at the University of Alabama. And we did it at the same time with the Freedom Riders. I therefore have had some experience in dealing with these kinds of problems." And: "As a member of the Cabinet, as a member of the National Security Council, I was involved in some decisions . . . I can't think of one decision, whether it's in connection with Panama, whether it was in connection with NATO, whether it was in connection with Latin America, whether it was in connection with the Congo. . . ." And on Vietnam: "When I spoke out in 1965. . . ." They were cheap shots, but they probably were scoring. McCarthy was reduced to telling the audience which Congressional committees he had served on, and in the experience-dropping game he ran a poor second—although he had been in public life for twenty years, longer than Kennedy, and was ten years older.

Also, while McCarthy was speaking to the television audience generally, Kennedy was zeroing in on Californians, and especially low-income Californians whose votes he needed in heavy majorities if he was to win impressively in the approaching primary. McCarthy talked generally of budget cuts necessary to pay for a war economy; Kennedy said: "We have cut 25 million from the Head Start program, which is going to mean a thousand students already in the Head Start program here in the State of California will be excluded." And: "We can fight for freedom 12,500 miles away, but we must do something to deal with the quality of life here. The property taxes, for instance, here in the State of California, which are astronomically high, the problems of poverty, all of these problems that have beset the American people, must receive our first priority." As a something-for-everybody sentence that one could hold its own; California property owners and the have-nots both could draw solace from it.

Finally in a discussion of low-income housing, Kennedy seized on a

McCarthy position and turned it around to make it sound as though the Minnesotan was being unrealistic, insensitive to Negro needs and—without precisely saying so—threatening the sanctuary of the white suburbs. By intent or inadvertence, it was precisely the sort of thing that had won Kennedy his image for ruthlessness, and although the press and McCarthyites were quick to spot it, Kennedy brought it off with a casualness that complemented his controlled demeanor throughout the debate. McCarthy himself, whose backers had been sure would take Kennedy apart, sat there, a bit uneasily, and took his lumps.

Kennedy had said he favored enlisting private enterprise to build housing in the big-city ghettos. McCarthy replied that "we have got to get to the suburbs, with this kind of housing, because some of the jobs are in the city and some jobs are being built there, but most of the employment now is in the beltline, outside of the cities, and I don't think we ought to perpetuate the ghetto, if we can help it, even by putting better houses or low-cost houses there . . . otherwise we are adopting a kind of apartheid in this country, a practical apartheid, and I see some industries are going to the ghetto, but some of the housing has got to go out of the ghetto, so there is a distribution of the races throughout the whole structure of our cities and into our rural areas. . . ."

Kennedy, again with an eye to California voters, fired back: "I am all in favor of moving people out of the ghettos, but we have 14 million Negroes here in the ghetto at the present time. We have here in the State of California a million Mexican-Americans whose poverty is even greater than any of the black people's. I mean, when you say you are going to take 10,000 black people and move them into Orange County, . . . [you] take them out where 40 percent of them don't have any jobs at all, that's what you are talking about. But if you are talking about hitting the problem in a major way, taking those people out, putting them in the suburbs where they can't afford the housing, where their children can't keep up with the schools, and where they don't have the schools for the jobs, it's just going to be catastrophic. I don't want to have them moved. Other groups have moved in the United States as they get the jobs and they get the training, then they can move themselves out into the other areas of the United States and will be accepted. They will find jobs and employment. But that does not exist, that's not the conditions that we are facing in this country at the present time."

The reference to moving 10,000 Negroes into Orange County was pure back-lash, inadvertent or otherwise. But McCarthy made no serious effort to challenge it then. Instead, a bit lamely, he replied: "There are an estimated 250,000 jobs available, but there aren't people within reach, and I thought when this question was first raised, that this was not your clear position, to concentrate that much on the ghetto."

Kennedy was back again, the voice of compassion. "I want to do things in the suburbs," he said, "but what I am saying is in order to meet the really hard-core heart of the problem, we have to face the fact that a lot of these people are going to live here [in the ghettos] for another several decades. And they can't live under the conditions that they are living under at the present time."

To the average viewer, the debate may have looked like a stalemate, and somewhat dull at that. But in terms of what each principal sought to get from it, Kennedy had the edge. Two debate performances by his late brother in 1960 seemed to set his style. As in the first Kennedy-Nixon debate, Robert Kennedy went on the offensive, displayed his grasp of the issues and demonstrated there wasn't as much difference between him and McCarthy as the age gap of ten years or appearances suggested. He had not, of course, dominated the debate the way John Kennedy did against Nixon, but he did help his cause. And his repeated technique of relating issues to the California electorate was reminiscent of John Kennedy's television debate with Humphrey in the 1960 West Virginia primary, in which the Massachusetts Senator more than matched Humphrey's discussion of national issues by focusing sharply on problems of primary concern to West Virginians—their poverty, their reliance on an inadequate food stamp plan, local unemployment and similar issues.

After the debate, Robert Kennedy's strategists were highly pleased. They felt that at the very least their man had held his own, and that the pesky question, "Why won't Bobby debate?" had been disposed of. Kennedy, in their view, had been forceful without losing his temper, without appearing "ruthless." McCarthy's followers for the most part were a bit let down. Not that they thought Kennedy had won; it was, rather, that many of them had persuaded themselves that if only McCarthy could get Kennedy in the same room with him, he would unmask the younger man as a lightweight. That had not happened. Instead of demolishing Kennedy, McCarthy had seemed at times to show no taste for the debate. In any event, he certainly had not made the most of an opportunity he had been demanding ever since Indiana. Questioned at the debate's close, Kennedy said he was "satisfied" with the way it had gone, McCarthy blandly called it "a kind of no-decision bout with three referees and sixteen-ounce gloves." Did he want a rematch? "No," he replied. "We'd get tired of each other."

One other question received little notice by most viewers. A panelist observed about midway through the hour that "many people think that the Middle East will be the area of the next great confrontation between East and West." He noted that Kennedy "this week proposed that we send fifty Phantom jets" to Israel, and he asked McCarthy if he agreed. The Minnesotan said he did. Kennedy himself was not asked to respond, and did not.

That night in San Francisco, another "gala" for Kennedy was held, with plenty of stars and celebrities. The Kennedys dropped by, but didn't stay long. The candidate, for all his energy, was a weary man. The long days and nights of campaigning had taken their toll; his face seemed puffed up at times and his hands always were, from the incessant handshaking and grabbing. Sometimes they bled at night, and trembled when he sat quietly on his plane or in his car. It was only an iron discipline that enabled him to maintain the pace. Eschewing the late bull sessions that go on endlessly among staff and newsmen on any campaign trip, he most often would return to his hotel room with Bill Barry or Fred Dutton, make a few phone calls, perhaps relax with a bottle or two of Heineken beer, his favorite, or one scotch, take a long hot bath and then retire.

The next morning, after taping a television interview in which he expressed the hope that after the California primary he and McCarthy could "somehow join and try to bring together" all the anti-Administration forces, Kennedy went to Sunday Mass with his wife. They attended St. Mary's, a small church a few blocks down Nob Hill from the Fairmont, at the edge of Chinatown. The Kennedys entered unobtrusively and sat to the right. Near the end of the Mass, for one of the rare times in public, both went to the altar and received Holy Communion. Outside later, Ethel insisted on plodding up the steep sidewalk to the Fairmount. She was buoyant about her husband's performance during the debate, and expressed surprise to a friend at McCarthy's. "Bob says McCarthy didn't do his homework," she said. "I can't get over it."

If McCarthy was derelict in his preparations, he was quick to add a postscript. In a statement Sunday he finally accused Kennedy, in his debate remark about moving Negroes into Orange County, of "scare tactics" that "could increase suspicion and mistrust among the races." The remark, McCarthy said, was "a crude distortion" of his position. "In past months I have proposed programs to enable Negroes to escape the oppressive conditions of the ghetto and find better lives outside the inner city, in the beltline and beyond, where most of the new jobs are," McCarthy said. And he noted that Kennedy "incidentally" was taking his campaign that very day into the same Orange County he had mentioned in the televised debate. In Watts, the Minnesotan was joined by Louis Lomax, a Negro writer and commentator, who endorsed him and said: "Senator Kennedy may have won some votes in racist Orange County by what he said last night, but he lost mine and I suspect thousands of others." And later in the day, in Bakersfield, McCarthy remarked, "Governor Reagan has been saying Senator Kennedy is talking more and more like him. I just didn't notice it until last night."

Kennedy meanwhile did go to Orange County, attended a Strawberry Festival in Garden Grove City Park and took six of his kids to Disneyland. They were mobbed by surprised tourists but managed to take a ride on "Pirates of the Caribbean." Because the crowds were so great, Kennedy promised to bring the children back another day. There would be, after all, some time off after the Tuesday primary.

On Monday, June 3, the final day of campaigning in California, Marat/Sade apparently was back at the Kennedy scheduling board. The candidate was to fly from Los Angeles up to San Francisco for a motorcade, down to Long Beach for another motorcade, to San Diego for a rally and back to Los Angeles—a distance of about 1,200 miles. But it was the last day, and it was important to hit all the major television outlets. The California State Poll, just out, gave Kennedy a 36 percent to 31 percent lead over McCarthy, with only 15 percent for the Lynch slate. But 18 percent were said to be undecided. If the pattern of undecideds going to McCarthy in other states were to follow in California, it could be too close for comfort. And if a last-minute bid by the pro-Administration forces to pass the Lynch slate off as a kind of McCarthy-Kennedy-Humphrey fusion group made any inroads, it could be worse. This last-day push was necessary, and besides, it was not in Kennedy to do less in the final hours.

In San Francisco, the caravan moved swiftly from the airport into down-town and on in the late morning to Chinatown, where it slowed for the cus-tomary handshaking and short speechmaking. The streets were lined three and four deep as Kennedy and his wife stood on the back seat and waved. As always, young boys and girls ran alongside the car and proprietors came out of their small shops to watch the spectacle. About the third block into Chinatown, just past an intersection, half a dozen sharp and loud shots rang out. Ethel jumped down and sat, hunched over, but her husband stood right where he was, continuing to do what he had been doing—waving, shaking hands. It was as though he had prepared himself for just such a moment, and to endure through it. The shots turned out to be firecrackers—big ones, like cherry bombs, whose report would make more lethal shots, from a .22 caliber revolver, say, sound like mere toy caps. The candidate looked down at Ethel, then around the car, and motioned to a friend running alongside to join her to steady her nerves.

For those who could not differentiate between the sound of firecrackers and the sound of gunfire, it had been a tense moment, but an understandable reaction. Others, such as those thousands of firearms enthusiasts in Califor-nia who flocked to shooting ranges daily to test their skills, would not have been confused. Certainly not one of them, a short, olive-skinned young man with black bushy hair who at the same hour was seen rapid-firing 300 or 400 rounds of .22 caliber bullets from a revolver at the San Gabriel Valley Gun Club in Duarte, a Los Angeles suburb. His manner of firing, in violation of range rules to shoot and pause, attracted the attention of Henry A. Carreon, a college student and playground director, and his friend David Montellano, who were practicing about five feet to his right. Carreon asked what kind of revolver he was using, and when the young man ignored him, he asked again, "Well, what kind of revolver is it?" And the young man finally answered, "An Iver Johnson."

The Kennedy motorcade through San Francisco went on to Fisherman's Wharf, where the candidate spoke at DiMaggio's Restaurant to a relatively small group of Italian-Americans. The reception was polite but hardly ecstatic. Kennedy enlivened the gathering with a few broadsides at McCarthy, whose belated charges of foul play in the debate had irritated him. "I have great ad-miration for the things he has done," Kennedy said, "but it seems to me the time to say something about someone is when he is there, face to face, rather than when he is five hundred miles away. He wanted to get me in a room with him. He got me in a room with him . . . and I thought he was very nice to me." Kennedy never had held McCarthy in very high regard, not from the beginning, when he doubted his ability to run a successful Presidential campaign, to arouse people. Now he had reason to have more respect for him as a politician, but he liked him less; and this latest long-distance sniping really got to him.

From San Francisco, the Kennedy party flew to Long Beach, where "a walk through the park" was scheduled. On arrival, 6,000 citizens jammed the park. "What am I going to do with them?" he asked Dutton. "Talk to them," his aide said with a grin. So Kennedy did, but he was weary and disjointed in his thoughts and words. He played with the crowd, laughed at himself in the middle of what were supposed to be serious observations. When a kid raised a sign too high, he said, "Down, down, down—that's it." And when the audience applauded he added, "It's nothing, ladies and gentlemen. That's just the beginning of what I would accomplish if I were elected President of the United States. There would be just all kinds of good things." He seemed for a moment to be losing control, to be throwing the crowd away on purpose. When he finished, and got back into the car, Dutton said to him, "You had a little trouble with some words that time." "I don't feel good," Kennedy replied. It was unusual for him to admit it, even if it were true, so Dutton watched him closely.

The motorcade went through Watts, taking side streets to make up time— and to avoid a too-frenzied reception that would scare off TV-watching white suburbanites. As the press bus careened down one residential street with hardly a soul out, Dick Drayne inquired of nobody in particular: "What is this? Are we going house hunting?"

Motorcading through the Los Angeles suburb of Venice, Kennedy asked Dutton to get him a bottle of ginger ale. The aide ran ahead to an outdoor stand, but all they had was a cola drink. He brought it back, but Kennedy couldn't drink it. Finally, Dutton bought a six-pack of ginger ale and Kennedy drank a bottle every few blocks, standing and shaking hands most of the way. Dutton got a second six-pack to take aboard the plane to San Diego. There, a surprisingly large crowd was at the airport and along the route to town. Kennedy told Dutton he was feeling better, but added: "The sooner we can go on, the better."

The rally was being held in an auditorium of the El Cortez Hotel. Some celebrities were there to entertain, including Kennedy friends Andy Williams and Rosemary Clooney. Kennedy decided to go right on. He raced through his remarks, ended abruptly, and then walked to the steps leading down from the stage. He sat on the top one, his face buried in his hands. The first impression was that it was an emotional thing, the last speech of the last preferential primary. Rafer Johnson, the Olympic decathlon champion, and Bill Barry hurriedly walked him down a corridor behind the stage to a dressing room about 40 feet away. He sat against the sink, Barry with him, while Rafer stood guard outside the door. Dutton went in, then Ethel. In a few minutes, the candidate came out, looking a little wan, but in control. He went back to the stage, addressed the crowd again, and finished the campaign the way he was supposed to. Turning to the reporters standing just to one side and below him at the platform, he said, "For the benefit of my friends on the left I want to add, as George Bernard Shaw once said, 'Some men see things as they are and say, "Why?" I dream things that never were and say, "Why not?" ' " And then he went over and sat with Ethel while Andy Williams sang for the crowd, and the Kennedys.

The candidate and party left the building in a wide freight elevator that got stuck a few times before discharging its weary passengers. The entertainers were returning to Los Angeles, and as Rosemary Clooney boarded the plane, Kennedy called for her to sit with him and Ethel. It was a subdued flight back, with small groups of friends and newsmen inquiring after the candidate's health, and a minimum of clowning around, especially for a Kennedy flight at the end of a long campaign. Tomorrow, June 4, would be a day off, at least until late afternoon when the first returns from the South Dakota primary would be coming in, and the sample precinct reports from California. The Kennedys decided to spend the night and the next day at John Frankenheimer's beach house in Malibu, with the six children. It would be their first free day together in some time, and they all looked forward to it.

Back at the Ambassador, the Kennedy headquarters hotel, an end-of-primary unwinding party got going in Dick Drayne's room. Some quiet drinking and conversation, with Dutton sprawled out on the floor and seven or eight others sitting around, grew by degrees into a minor and nominally raucous blast. John Hart of CBS put the gathering in the right mood with an ear-splitting and side-splitting version of the Reverend Billy Graham at the peak of his persuasive powers, and before long the drinkfest had matured into a songfest. When an irate hotel guest appeared at the open doorway, obviously not appreciative of Drayne's Revelers, the group trekked up to the Kennedys' unoccupied suite and escalated their antics. For another hour or so, we drank the Senator's liquor, paraded around his living room to a medley of robust marches and generally ensured ourselves some of the better hangovers of the campaign. But the next day would be an easy one, especially for the press. There would be nothing of note going on until after the polls closed, and then, no matter who won, just the same routine victory speeches and press conferences. If you'd seen one election night, you'd seen them all.

CHAPTER EIGHT

Triumph—and Despair

THE LAST DAYS

EIGHTY DAYS OF constant, punishing street campaigning that had sapped the strength and stamina from the slight frame of Robert F. Kennedy, and had brought him to the brink of collapse in San Diego the night before, were behind him now. And so on the morning of the final preferential primary, in the sumptuous Malibu beach house turned over to him by John Frankenheimer, the candidate surrendered to the luxury of sleeping late. He rose restored shortly before eleven o'clock and telephoned his weary lieutenant, Dutton, back at the Ambassador. The songfest in the Kennedys' suite had left Dutton more in need of additional sleep than of athletics at the beach, so on this lazy and most welcome of days on the campaign trail, he talked briefly with the candidate, promised to drive out to Malibu in the early afternoon, and went back to bed.

After lunch with wife Ethel, six of their ten children, and author Theodore H. White, Kennedy took to the beach with the kids. It was chilly and overcast but he plunged into the waves anyway, just as he had done on the Oregon coast that day in the late stages of that lamented campaign. Suddenly, he spied twelve-year-old David, pulled down by an undertow. He dove in and came up with the boy, and a red bruise on his own forehead. It had been a close call for the boy, and in most other families it would have been cause to end the day's outdoor activity for all the children. But not in the Kennedy family. After David's close call, there was more frolicking in the Frankenheimers' pool and, when brother Teddy, Goodwin and Dutton all had arrived, some political talk. In the CBS News spot check of 400 sample precincts on primary days, half is taken at around noon and the other half about an hour or so before the polls close. The network by about 3 o'clock phoned the first half of the sample to Mankiewicz, who passed it on to the other insiders, and the word was good—about 49 percent. Or at least the aides thought it was good. The candidate himself was uncertain; after the Oregon defeat, it would take more than an eyelash victory, maybe a majority, to persuade the Daleys and the Tates to come around. But the others were encouraged, so eventually he joined the mood of optimism. He spoke of the bitterness of McCarthy, but said he did not believe the differences between him and his foe on the primary trail were too great for them to get together. After all, he recalled, the feelings ran pretty high between his brother and Humphrey in 1960, and Humphrey had supported John Kennedy in the end. On this hopeful note, the aides finally disbanded and Kennedy napped.

As the sunless afternoon limped on, the curious Kennedy mix of political ditchdiggers and dilettantes began to gather. The Senator himself was dressed by about 6:30 and impatient to head in for the Ambassador. But Ethel wasn't ready. At Kennedy's urging, Frankenheimer wheeled out his big and flashy car and chauffeured the candidate and Dutton into town. A taut, nervous, high-strung Hollywood stereotype of the Hollywood director, he drove a round-about route in a manner that had Kennedy on edge and pressing him to keep to the speed limit.

When they arrived at the headquarters hotel, shortly after 7 P.M., the festive election-night mood already was in command. The fifth-floor Royal Suite, which has an entrance hallway with a huge living room to the right and two bedrooms and bath to the left, already had been put under siege by nearly 100 assorted Democratic functionaries, privileged reporters and other lesser hangers-on. They were crowded into the living room, where television sets and a bar already were servicing their needs. Kennedy, Dutton and Frankenheimer went into the bedroom section, where a much smaller and more select group was gathering. Then or shortly afterward, Ted Kennedy was there, and Ted Sorensen, Dick Goodwin, Steve and Jean Smith, Pat Lawford, reporters Hays Gorey, Dick Harwood and Jimmy Breslin, photographer Stan Tretick, Jesse Unruh, and Ethel, bright-eyed and excited.

Early on, Dutton called South Dakota, where a less important but noteworthy primary had been held that day: the word was very good, possibly a majority. "You want to hear about the Indians?" Kennedy asked those intimates who always kidded him about "the suicide rate." "We've got an Indian precinct in from South Dakota: 878 for Kennedy, 2 for Johnson, 9 for McCarthy. How about that?" (The final statewide results gave Kennedy 49.7 percent to 29.9 percent for native-son Humphrey and 20.4 percent for McCarthy.) The candidate was boosted by the prospect of two victories on one night and he personally took the phone and thanked his South Dakota manager, Bill Dougherty.

As the California polls closed at 8 o'clock, the Embassy Room already was packed with Kennedy celebrants. The mostly young crowd, boys wearing Kennedy straw hats and girls in white blouses, blue skirts and red Kennedy sashes, listened and clapped to a rock band while others watched the first speculative television reports over one of the six color sets placed around the room. Roger Mudd stood before a camera with a group of teeny-boppers around him, waiting for his cue to report the early vote projections of his network. Almost from the outset, the CBS calculations had Kennedy winning, and comfortably. But NBC's sample precincts were slow reporting, and the network held back. Chuck Quinn of NBC waited inconspicuously across the room for the projection, with Larry O'Brien in tow. At about 8:25, the first reports of actual returns had given McCarthy 49 percent, Kennedy 40 percent; but these were all from outside Los Angeles County, which with 38 percent of the state's total registration was expected to be a Kennedy stronghold. Quinn and NBC held off, but O'Brien had no doubts about how it was going.

In Oregon, where there had been a large undecided vote in the final polls, that vote had gone almost exclusively to McCarthy. This time it was different, O'Brien insisted to one reporter. "I think we broke the pattern of the undecideds," he said. "I think the debate did it. Every survey of ours indicated two out of three came over to Bob after the debate. Our canvassers found that on Sunday and Monday."

The cordial Irishman from Massachusetts, accustomed to winning elections, was back on the track and quietly pleased about it, in the manner of the professional who expects and is expected to do his job well. He talked of plans to leave the next morning for New York, to help steer the campaign for Kennedy convention delegates in the June 18 primary there. Although Kennedy as the ranking New York Democrat was expected to win the bulk of them, McCarthyites were challenging in most of the forty-one Congressional districts.

"It will be tough," O'Brien said. Although neither Kennedy nor McCarthy would be on the ballot in any preferential test, each would have to campaign to establish identification with their delegate candidates. McCarthy was to go up to Washington State on a delegate hunt and then head for New York; Kennedy was to enjoy two or three days of rest in the California sun, then plunge in too.

But for the time being, the focus remained here in California, on the second-largest bloc of votes. In years past, before television's technology and huge audience preempted tradition, it was customary for the candidates to appear first before their party workers, and the press, to give a statement of victory or of concession. But now "the tube" swept all that away, and the candidates elected to take the route of "maximum exposure"—that most desirable of conditions in this most political of all civilizations. First it was McCarthy, interviewed by CBS' David Schoumacher at the Beverly Hilton Hotel, his headquarters several miles from the Ambassador. The Minnesotan did not challenge the projections, but characteristically sought to dismiss California's significance.

"We made our real test in Oregon," where, he said, there were no minority blocs to distort the results as in California—a studied effort to downgrade Kennedy's strong appeal to Negroes and Mexican-Americans. Any Democrat would do well among these voters against a Republican opponent like Nixon in November, McCarthy argued; the important test was among the independents: "We're demonstrating what we said we would—that I can get votes no other Democrat can get."

And then he added: "We neglected California to go into Oregon." It was a transparent alibi, reminiscent of his earlier attempts to write off the importance of his losses in Indiana and Nebraska. McCarthy had spent all of the final week in California, had shuttled there from Oregon several times during that primary campaign, and had his largest and most professional of all organizations in California. It was true he had raised Oregon's priority as he and aides sensed a chance for an upset, but all along California had been regarded as the key primary test.

Now it was Kennedy's turn on TV, but he was apprehensive about making any overly optimistic claims, especially when the networks were not in agreement and when the Los Angeles County vote, delayed in a comedy of computer errors, had not been fed in. The television people were pressing him; because of the three-hour time difference if he didn't appear in the next few minutes he would lose the New York audience; or he would lose out in Chicago, where it was two hours later. Still the candidate stalled; it would be better not to say anything to the television audience than to make extravagant claims that might look foolish later. But privately, there were things that could be done to plan ahead, assuming a comfortable California victory. Kennedy phoned O'Donnell back East to discuss a major strategy meeting for which O'Donnell was flying out to California the next morning. Kenny told him it looked as though he would be nominated, and Kennedy replied, "I think I may." O'Donnell said later: "He had arrived. He had won the biggest state in the Union—not as Jack Kennedy's brother, not as Bobby Kennedy, but as Robert Kennedy."

The candidate also tried to call Al Lowenstein, the original dump-Johnson apostle still supporting McCarthy, in New York shortly before 11 o'clock; he wanted to talk about how he now could pull in the McCarthy supporters, and about how the New York primary was shaping up. Unruh was dispatched downstairs to speak to the ballroom enthusiasts and keep them at bay.

Finally, Kennedy could hold off the insistent television people no longer. Characteristically, they had laid siege to the suite with logistical overkill. A mini-skirted girl armed with a walkie-talkie hovered outside the suite ready to report the candidate's departure; producers and assistant producers worked on those Kennedy aides who might best be able to spring the candidate. When the Senator moved out into the corridor at one point and a group of local reporters rushed up to him, Dutton led him off. If he were going to talk, he might as well talk to the country over television. The candidate went down one floor for an interview with Sander Vanocur of NBC, expressed his pleasure with the early California indications and renewed his pitch to McCarthy and his backers to join him if Humphrey was to be denied the nomination. Then he returned to his own floor for a similar session with Mudd in the CBS room. Kennedy was in a good humor now. The latest returns from precincts exclusive of Los Angeles showed him creeping up, with 41 percent to McCarthy's 46. The city surely would push him well ahead. He told Mudd that as far as McCarthy was concerned, he hoped "we could come together," that "perhaps we could unite now that the primaries are finished . . . and try to accomplish what we all started out to accomplish, which was for a cause, for a purpose, not for an individual."

Kennedy was being his most affable, mindful of his "ruthless" image. But thanks to Mudd's phrasing of questions, the interview soon became one of the most authentic appearances Kennedy ever had made on television—capturing his wry sense of humor as it seldom had been seen in public.

"Well, it appears though, doesn't it," Mudd asked, "that you're not going to be able to *shake* Eugene McCarthy?" Kennedy seemed to wince at the word. "I think, really, it's up to him," he said, backing away from the sharp question. "I mean, he's going to have to make that decision himself. . . ."

The subject turned to Vice President Humphrey. Asked Mudd: "You have no way, now, between California and Chicago, to *draw the Vice President into a fight?*"

Again Kennedy winced. "No, I . . . Do I have to put it that way?" At every question, "ruthlessness" seemed to be rearing its ugly head, and there he was, on his very gentlest behavior. What he would like, Kennedy said, would be to have a public discussion with Humphrey around the country. "I don't think that the policies that he espouses would be successful in the country and I don't think they'd be successful with the Democratic Party and I think all the primaries have indicated that," he said.

"Well," interjected Mudd, "are you saying, Senator, that if the Democratic Party nominates the Vice President, *it will be cutting its own throat* in November?" Again, the wince, this time accompanied with a grin. "Well, again," Kennedy told Mudd, "you use those expressions. I think that the Democratic Party would be making a very bad mistake to ignore the wishes of the people and ignore these primaries."

There was a question on the possibilities of a Humphrey-Kennedy ticket, which Kennedy turned away softly. Then he allowed himself a slight poke at Humphrey for suggesting, Kennedy said, that if Johnson decided after all "that he wants to run, that he would then step aside. That I don't understand."

Mudd was right there to put it into the vernacular. "You felt that was fairly *shoddy politics,* I take it?"

Kennedy reacted like a man who had just been given a hotfoot. "Well, again, I . . . Roger—I, you know, I don't think . . ." he couldn't find the words to adequately—yet benignly—register his protest. Finally he just went on and talked some more politics. But Mudd was not through.

"Are some of the delegates that are listed as leaning or even committed to the Vice President," he asked, always in good humor, "are they *squeezable?* Are they solid?"

Kennedy could no longer restrain himself. "Roger, your language!" he said in mock horror. "I don't like either of those expressions."

Mudd (laughing): "Well, that . . . you—isn't that the way you talk about it?"

Kennedy (protesting his innocence, but laughing too): "No, I don't go that far, I don't, I don't."

Mudd: "Well, I . . ."

Kennedy: "Probably somebody else does."

By now the interview was a marvelous shambles. Mudd asked one more quick question, then said: "Thank you. I'll work on my language for the next time, Senator." Kennedy grinned. "Thank you," he replied.

The candidate, still laughing at the session with Mudd, was whisked through a bathroom into the nearby ABC room, where newscaster Robert Clark interviewed him, and then to a short radio interview with Dan Blackburn of Metromedia News. By this time, the crowd downstairs was getting unruly, and Mankiewicz had come up to urge the candidate to make his victory statement. NBC finally had produced an optimistic report for Kennedy and the statistics were beginning to level off with Kennedy safely ahead. "If you're going down there," one aide told him, "you'd better do it now, because you're not going to get any better indication." Unruh, the California sage, agreed, and said he was safely ahead.

Mankiewicz, Dutton and Kennedy retired to the inner bedroom to discuss what he would say. "We talked about California and gloated a little," Mankiewicz recalled later. As Kennedy walked around the room, the two aides threw out names of those supporters Kennedy ought to thank, and the points he ought to make—about winning in an urban state like California and a farm state like South Dakota in one day; about the evidence that the people didn't want a continuation of the Johnson Administration; about the need of McCarthy supporters to join with him to prevent that from happening. Finally, the three men walked out of the bedroom, past the privileged well-wishers and into the corridor. Kennedy asked Goodwin to call Lowenstein again, to tell him he would phone immediately after he returned from the victory speech. The delegate primary in New York was to be the first order of business at the next day's strategy meeting, and Kennedy was anxious about the situation there. Goodwin placed the call, and Lowenstein stood by in his Long Island home.

The revelry continued and the guests in the Royal Suite turned to the television sets to watch the triumphant candidate's entry into the ballroom below. Some thought they might go down, but the prospect of the crush, one just like another in a Kennedy campaign, was too much to endure for a routine election-night obligation they had witnessed so many times before.

It was about fifteen minutes before midnight when Kennedy came down the service elevator and through the hotel kitchen, buoyant now over what he believed to be his solid comeback victory. He shook hands with kitchen workers and then walked out through the narrow, gray corridor that linked the kitchen to the Embassy Room. A television photographer toting a portable camera on his shoulder backed out a few feet in front of the smiling Senator, natty in a dark blue suit with blue and white striped tie. A Los Angeles fireman standing there, Harold L. Burba, said later that he saw walking behind Kennedy a young man with bushy hair holding a rolled-up poster who "appeared to be looking all around him instead of looking ahead as all the other members [of the party] were . . . he seemed to look at me as he passed by. But then he looked on around, but he was looking all over all the areas as he passed by." Kennedy, ahead, grinned when he saw Bob Healy of the Boston *Globe* and me standing there, took congratulations from us and invited us to his victory party later, at Pierre Salinger's discothèque The Factory. Not a man given to open enthusiasm, he obviously was in a puckish mood, as elated as either of us had seen him anytime during the campaign. I ribbed him about the Roger Mudd television interview, telling him I thought he had been "very ruthful" on the show. The Senator laughed heartily, took a few steps down the corridor, then turned and said, smiling broadly: "I'm getting better all the time."

That was the last private moment, the last characteristic self-deprecation. Then he was out in the hot, super-lighted ballroom, being screamed at, touched, pulled by his joyous, celebrating California supporters. As the band blared campaign songs, Kennedy—with Ethel wedged in at his side, little-girlish in orange and white mini-dress and white stockings—mounted the small platform. Again the screams and the cheers went up as those in the rear saw the familiar, tousled head for the first time. Photographers scrambled behind him, clawing at each other in the peculiarly masochistic style that is their professional trademark, recording the moment of triumph from every conceivable angle. Ethel waved and then turned and gave a friend off to the side the mild look of resignation that frequents those who are trapped repeatedly in public acclaim. Dutton, standing at the rear of the platform, said to me of our songfest the night before: "When I told the Senator what went on in his suite last night, he was like a little boy who missed out on something."

But Robert Kennedy was not going to miss out tonight. The party at Salinger's discothèque was for him and invited friends, and he and Ethel seldom were in better spirits for a party. But first the responsibilities of being a winner had to be satisfied.

The tumult faded to a buzz and Kennedy spoke into the battery of microphones, conveying at once his playful mood, his sense of revival after the defeat of just one week earlier. He congratulated Don Drysdale, the Los Angeles Dodgers' pitcher who had just broken a major-league record. "He pitched his sixth straight shutout tonight," Kennedy said, "and I hope that we have as good fortune in our campaign." He thanked Steve Smith, who he said was "ruthless" in running his campaign, and those in California who had been particularly instrumental in his victory: Cesar Chavez; Jesse Unruh; Paul Schrade, a California United Auto Workers official; Rafer Johnson; and Roosevelt Grier, the huge Los Angeles Rams All-Pro tackle of whom Kennedy remarked: "Rosey Grier said he'd take care of anybody who didn't vote for me." The crowd laughed appreciatively at the observation, which was to take on grim poignancy only moments later. "I want to express my thanks to my dog, Freckles," he went on. "I'm not doing this in the order of importance. I also want to thank my wife, Ethel." More laughs by those who were so ready to laugh.

Then Kennedy began the burden of his victory talk. He expressed his gratitude especially to those in farm areas in both California and South Dakota who had voted for him. "Here is the most urban state of any of the states of our Union, South Dakota the most rural of any of the states of our Union," he said.

"We were able to win them both. I think that we can end the divisions within the United States.

"What I think is quite clear [is] that we can work together in the last analysis, and that what has been going on within the United States over a period of the last three years—the divisions, the violence, the disenchantment with our society; the divisions, whether it's between blacks and whites, between the poor and the more affluent, or between age groups or on the war in Vietnam—is that we can start to work together. We are a great country, an unselfish country and a compassionate country. I intend to make that my basis for running."

The temperature in the ballroom was oppressive now, with the jammed-in bodies and the television lights compounding it all, and he wanted to get off to the party. "I'll just take a moment more of your time because everybody must be dying from the heat," Kennedy said. "What these primaries have indicated and all of the party caucuses have indicated, whether they occurred in Colorado or Idaho or Iowa—wherever they occurred—was that people in the Democratic Party and the people of the United States want a change. And that change can come about only if those who are delegates in Chicago recognize the importance of what has happened here in the State of California, what has happened in South Dakota, what's happened in New Hampshire, what happened across the country. The country wants to move in a different direction, we want to deal with our own problems within our country and we want peace in Vietnam."

Kennedy then reiterated his bid for reconciliation within the anti-Administration camp, congratulating McCarthy and his followers for their efforts from New Hampshire and through California. "The fact is," he stressed, "that all of us are involved in this great effort. It is a great effort not only on the part of the Democratic Party, it is a great effort on behalf of the United States, on behalf of our own people, on behalf of mankind around the globe." He spoke now with the kind of quiet, pleading, impatient intensity of the teacher of small boys who had trouble getting across the obvious.

"I would hope now that the California primary is finished," he went on, "now that the primary is over, that we could concentrate on having a dialogue— or a debate, I hope—between the Vice President and perhaps myself on what direction we want to go in; what we are going to do in the rural areas of our country, what we are going to do with those who still suffer within the United States from hunger, what we're going to do around the rest of the globe, and whether we're going to continue the policies that have been so unsuccessful in Vietnam—of American troops and American Marines carrying the major burden of that conflict. I do not want this and I think we can move in a different direction."

He was more than ready to go now, but did not want to shortchange them. He had run so many campaigns, and he knew what these people had done for him. "I thank all of you who made this possible this evening," he said, and the V-for-Victory fingers were flashed up from the beaming young crowd. "All of the effort that you made and all of the people whose places I haven't been to, but who made or did all of the work at the precinct level, got out the vote, did all of the efforts—brought forth all of the efforts required. I was a campaign manager eight years ago and I know what a difference that kind of effort and that kind of commitment can make. My thanks to all of you, and on to Chicago."

They were waiting to let go, and now they did, cheering and applauding and pushing toward him. He reached down and shook hands with some of them and touched some outstretched fingers, as the photographers sprang into action again, working their cameras or moving off the platform to record his exit. The original plan had been for Kennedy to go out through the crowd to another reception on a lower level of the hotel. But there was closed-circuit television down there and that group had just seen and heard the candidate. When Bill Barry came up to Dutton in the crush and told him the functionaries below wanted Kennedy, Dutton demurred. It would just be a repeat performance. Besides, it was already past midnight; the writing newsmen as usual were pushing for a press conference, and the shortest route to the press room was the way Kennedy had come in, through the kitchen corridor and pantry area. In Oregon, where he had lost, he bypassed that stop, but in Indiana and Nebraska, where he had won, the trip to the press room was his first order of business after the speech to his workers. Barry and Dutton discussed the matter another moment and then, as Kennedy concluded and started out, Barry moved up and told him: "No, it's been changed. We're going this way." "Okay," Kennedy said, still swamped by admirers.

Dutton and Barry, expecting the candidate to leave by way of side steps that led onto the ballroom floor and then directly out into the kitchen corridor through swinging doors to the left, moved in that direction to clear a path for him to the press conference. But Kennedy, with Ethel slightly behind, was boxed in by reaching, screaming well-wishers chanting "We want Bobby! We want Bobby!" Instead of following his two aides, he started off the platform, led by the assistant maître d'hôtel, Karl Uecker, through a back exitway into the dark corridor. Barry looked back, saw the new route, and raced over. Kennedy already had started out and along the corridor, through an open doorway into a well-lighted area. Walking ahead of him about 20 feet toward the press room, I noticed two well-dressed women sitting nonchalantly in deck chairs in a little alcove to the right of the doorway. Their blase manner, in contrast with the excitement and the usual Kennedy jumpers and screamers, caused me to note mentally that they must be wives of California politicians, used to this sort of thing. It was after all, old hat for those who had been involved in Presidential campaigns over the years.

I walked on into the pantry area, past a large ice cube-making machine that went from floor to ceiling on the right, and two stainless steel steam tables on the left that narrowed the passageway to about six feet at one point. I noticed nobody else in particular, just the usual kitchen hands. But I did not look to my right as I passed the ice machine, where a short, dark young man waited, standing on a low tray-stacker. Shortly before midnight, the young man had asked a kitchen helper, Jesus Perez, three or four times: "Is Kennedy coming through here?"

Kennedy moved down now to where the ice machine and steam table narrowed the corridor—precisely where only minutes before, on his way to the ballroom, he had stood and joked about his appearance on Roger Mudd's show. He was only about 30 feet from the entrance to the Colonial Room, the makeshift press room that was his destination. Members of the kitchen staff lined his path to see him or shake his hand, and the Senator reached over the steam table to greet a female kitchen worker. Uecker, his left hand holding Kennedy's right, was guiding the Senator through, sometimes having to let go as Kennedy shook hands with an admirer, then taking it again as he led the way through the pantry area. Mrs. Kennedy still was behind, and Kennedy aides were clustered around. Barry had practically caught up and Dutton was not far behind. It was about 12:13 A.M., Pacific Time. A Mutual radio reporter, Andrew West, with his tape recorder running, asked Kennedy how he was going to counter Vice President Humphrey's delegate strength. Kennedy replied: "It just goes back to the struggle for it . . ."

He never finished. He turned routinely as if to look for his wife, and suddenly the dark young man stepped down from the tray-stacker, raised his right hand high over the surrounding aides and fired a stub-nosed revolver at Kennedy's head, at close range—probably no more than a few feet. There was, according to most accounts, a single shot—a quick "pop" that hardly sounded lethal, like a firecracker or a boy's cap gun—then a pause, and then a rapid volley: "pop-pop-pop-pop-pop-pop-pop." Some witnesses, however, said there were two distinct shots before the pause and the volley. Many of those who heard the lightning shots said later they thought they really were firecrackers, like those set off in San Francisco's Chinatown the day before. The witnesses differed in their recollections of how many shots there were; hardly anybody thought at the time there had been eight.

It was a paralyzing moment, except for Kennedy, who threw his hands up toward his face, then staggered back, falling to the gray concrete floor on his back—his eyes open, his arms over his head, his feet apart, about two feet from the side of the ice machine. He was alive, but grievously wounded; blood flowed from behind his right ear. In back of him, others were hit and fell.

It was all a blur of madness, with each witness to those terrible moments remembering the details from his own experience. In all the frenzy, it was not surprising that there were discrepancies, some of them substantial, in the various accounts. Uecker, the assistant maître d'hôtel, provided this account afterward to the Grand Jury:

". . . I was trying to get as fast through the kitchen area there, through the pantry, as I could. And while I was holding his hand, I was turning to my right towards—to the Colonial Room where the press room was. At the time something rushed on my right side. I—at that time I didn't recognize what it was, and I saw some paper flying. I don't even remember what it was, paper or white pieces of things. Then I heard the first shot and the second shot right after that, and Mr. Kennedy fell out of my hand. I lost his hand. I looked for him, and I saw him falling down. And I turned around again, and I saw the man—right standing next to me.

"The arm [he] was holding the gun in, [I] pushed the arm down on towards the steam heater, and my right arm I took around his neck as tight as I could, and pressing him against the steam heater. In the meantime, somebody else came behind me and pushed me against the steam heater. The guy in front of me couldn't get loose. While I was holding the hand where he had the gun in, I was trying to get the point of the gun as far as I could away from the part where Mr. Kennedy was laying. From the left side, I was trying to push the gun away to the right side where I didn't see too many people, while he was still shooting . . .

"I couldn't hear too clearly, but I thought it was five or six [shots], but I was hitting his hand on the steam heater as hard as I could, with my left hand, I had him right here on the wrist, and hitting my left hand on the heater to get rid of the gun . . . I was standing there and he was shooting, and I could feel when he was turning his hand towards the crowd, that's why I pushed all over the steam table as far as I could, to almost to the end of the steam table . . . I kept him all the time . . . until the gun was—must have been empty, and I was still hollering, 'Get his gun, get his gun!' And finally, after I didn't hear no more shots, many people . . . was around by that time; I don't know how many people because I was facing the wall. The people behind me were pushing against the steam table and I had—still was holding on to him. And then I saw some hands coming over, reaching for the gun. I don't know who it was."

Uecker testified that the gun had gone off twice and "I must have grabbed the arm by the third shot . . . While I was holding the right arm from the shooter, I looked to the left, and while he was shooting still, I saw some more people falling down."

Another hotel employee, Edward Minasian, a banquet captain, told the Grand Jury he was walking along with Uecker just ahead of Kennedy when "out of the side vision, my peripheral vision, I noticed someone dart out from this area, dart out and lean against the steam table. And I saw a hand extended with a revolver, and I saw the explosion of the cartridges out of the—out of a revolver." The man was holding the gun at shoulder height approximately three feet from Kennedy, Minasian testified, when he fired two shots.

"It was a bang-bang cadence, and after the second shot, why, as I said, I saw the flash of the cartridges being discharged, and immediately there were several other people in that area behind the senator, and I just pushed into Karl Uecker . . . We both made an attempt to get at the hand holding the gun, and we had him—I was down low, pushing up against him. And at that same time I turned to my left and I saw—well, there were some more wild type firing which was more rapid fire than the first two, as they were struggling for the gun. I'm sure that's why the gun was going off."

Bill Barry, who was not called before the Grand Jury, later gave his account:

"I had just arrived at the Senator's right, parallel with him. I heard shots. They didn't sound like shots, and I know what they sound like. I broke into the clear—I realized later it was because everybody was ducking. I swung left and nobody was between me and the man. He was standing with the gun when I hit him. I hit him with the edge of my hand. He dropped the gun on the table. The Senator was falling away from me. The gun was only twelve inches away when I hit him. I swung him around and hit him twice with my fist, two rights high on the cheek. I hit him two blows I think I would have floored Rosey Grier with. When I hit him the second time I knew I was going to kill him. So I put a stranglehold on him. Then I wanted to go back to the Senator. Rosey and Rafer were there. I called to Rosey to take him. At this time, people started hitting him. A guy climbed up on the table and began kicking at him. When Rosey came up, I said, 'Rosey, take this guy. Get this guy off in a corner where people can't hit him.' But as I was transferring him, he squirmed and got the gun again. Then others came up and grabbed his arm. I went over to the Senator, took off my jacket and put it under his head. I knew immediately it was a .22, a small caliber, so I hoped it wouldn't be so bad, but then I saw the hole in the Senator's head, and I knew."

Another eyewitness, Vincent DiPierro, a nineteen-year-old college freshman working as a waiter that night, told the Grand Jury later of seeing the assailant step down from the tray-stacker "as though he was clutching his stomach, as though somebody had elbowed him . . . and when I went to turn, the next thing I saw was him holding the gun . . . he kind of moved around Mr. Uecker. He kind of motioned around him and stuck the gun straight out. And nobody could move. It was—you were just frozen. You didn't know what to do. And then I saw the first powdering or plastering. When he pulled the trigger, the first shot, Mr. Kennedy fell down." Kennedy, the boy said, "was on his way, falling—he was falling down. The first shot, he kind of reared back very, very sharply . . . Both hands went up . . . the suspect turned almost immediately, and after all the shots were fired, he was trying to escape. He tried very, very hard to get away. But Mr. Rafer Johnson—and by this time Mr. Grier and Mr. Uecker were all holding him against the heater . . . the stainless steel table. And people were trying—were hitting him and cursing at him, and it was utter confusion. I mean, everyone was trying to kill him . . ."

Others also froze for an excruciating moment, and only when the deed was done did the scene come alive to them—wildly, deafeningly, angrily. The first thing most recall seeing was the desperate wrestling with the man by Grier, Johnson, author George Plimpton, and others who grabbed at him as he held the gun over his head. The huge figure of Grier dominated as the crush of bodies hurled the small man over to the left, onto the end of the steam table, close by where Kennedy lay. Hands grappled frantically for the gun as shrieks of grief, shock, desperation and disbelief filled the corridor:

"My God! He's been shot! He's been shot! Get a doctor! Get the gun! Get the gun! Kill him! Kill the bastard! No, don't kill him! Don't kill this one! Oh, my God, they've shot Kennedy!"

There were perhaps thirty persons in the room when it happened, and now the door to the press room was swung open and reporters, having heard the shots, raced in, catching the scene of the struggle with the gunman to their right, and a crowd of photographers and others around Kennedy straight ahead. The cameramen went to work with their usual blunt proficiency, to the point where they seemed for a time to be smothering the fallen Senator. "Get back! Get back! Give him air!" shouted Drayne and McDonald, the other press aide. But still they pressed in, cameras grinding and snapping. A woman photographer, on the edge of hysteria, could not work. "You can have it! You can have it!" she yelled to her male colleagues. Then she began tugging at one of them, trying to pull him back. He turned on her and said, "Get away! This is history!" And he went on with it.

Other photographers and reporters jumped up on the steam table, next to where the gunman was pinned, and looked down on Kennedy. Somebody had wiped his face of blood now, but still it trickled from the back of his head. His eyes were open and staring, but he still seemed conscious. His right hand was bloody and it lay on his stomach, leading some to believe he had been wounded there, which he had not. A young kitchen boy, Juan Romero, knelt at his side, took a set of rosary beads from his shirt pocket, placed it in Kennedy's hand, and prayed quietly. Kennedy's tie had been loosened now, his shirt opened to the waist, and his shoes removed. McDonald much later was seen wandering aimlessly about, still clutching the shoes.

Incredibly, Grier and the others still had not wrestled the gun from the man. Indeed, as he struggled wildly, his dark eyes wide, the throng of captors—even the gigantic Grier—seemed to have trouble restraining him. Few had accurately counted the shots, and there was the feeling that at any moment he might squirm loose and start firing again. Finally a host of black and white hands pinned the man's arm to the steam table and stretched it out, so that the revolver faced a blank wall away from the crowd. But still they couldn't get it from him. They chopped and flailed and a man standing on the steam table even kicked at the fingers, but they were locked on as if hand and gun were one. West, the Mutual reporter caught up in it all, in his anxiety crossed the line between reporter and participant in a narrative that captured the frenzy and the emotion in the horror-filled room:

"I am right here and Rafer Johnson has hold of the man who apparently has fired the shot! He has fired the shot. . . . He still has the gun! The gun is pointed at me right at this moment! I hope they can get the gun out of his hand. Be very careful. Get the gun . . . get the gun . . . get the gun . . . stay away from the gun . . . his hand is frozen . . . get his thumb . . . get his thumb . . . get his thumb . . . get his thumb . . . get his thumb . . . take hold of his thumb . . . and break it if you have to . . . get his thumb. Get away from the barrel! Get away from the barrel, man! Look out for the gun! Okay, all right. That's it, Rafer, get it! Get the gun, Rafer! Okay, now hold on to the gun. Hold onto him. Hold on to him. Ladies and gentlemen, they have the gun away from the man. . . ."

Someone had finally broken the man's grip, and the gun fell. Grier, Johnson and others reached for it, and Johnson got it. "Don't kill him! Don't kill him!" the cries went up again. Grier held the man in a headlock, and now that the gun was gone, others came up and cursed and punched at the man. "You monster!

You monster! You'll die for this!" a young kitchen aide standing on the steam table shouted at him. The assailant's dark eyes shot out in panic, but he said nothing. He was bushy-haired and his complexion led many to believe at first that he was a Mexican or Mexican-American. But that didn't make sense. Kennedy was so popular among these people. But none of it made sense.

Several minutes had passed by now, and still the police had not come, nor did there seem to be a doctor. Mrs. Kennedy, who had been pulled back to safety when the gunfire broke out, now was brought forward by a friend. She came up behind where her husband lay, but because of the crowd and the bedlam she did not see him until she was at his head.

"Oh, my God," she said, in a half-whisper. She moved quickly to his side, quietly, tenderly, in sharp and punishing contrast to the pandemonium around her. She stepped across him, knelt and took his hand. He turned his head and seemed to recognize her. She knelt there, in her orange and white party dress, on both knees, stooped low over the cold concrete floor, whispering to him, stroking his bare chest and brow.

Still the photographers pressed in. Drayne and McDonald tried to get them back to little avail. Unruh, exploding, began to push them, at one point mounting the steam table and shouting, "Get the hell out of here!" Some backed off, and he calmed down. But others remained. Finally, Mrs. Kennedy, her face full of anguish and pleading, but without tears and under control, rose and turned to those on the steam table. "Please go, please go," she said plaintively. "Give him room to breathe." Then she returned to her husband's side, and some of the photographers gave way. Mankiewicz joined in the effort to get the rest back, and so did Barry. Still there were no police. An officer finally did come into the long narrow room, but he was a private security officer. He drew his handgun and Barry, seeing it, told him sharply: "No. Put it away. Get these people the hell out of here." The officer put the pistol back in its holster.

For McDonald, the young press aide, the scene finally was too much. He began to sob, Barry turned to him and said: "Stop crying and do your job." It was said not in a bitter tone—more as an attempt to ward off emotionalism that could interfere with the task at hand. McDonald went about clearing the area.

Grier and the others now had pulled the assailant down the countertop about 15 or 20 feet from where Kennedy lay. Some were still pummeling him with fists and yelling obscenities. Others were coming up and trying to strike him. Unruh called out to no one and everyone: "I want him alive. If anything happens to this one, you answer to me. We don't want another Oswald!" Grier, still holding the man in a headlock, pushed away others who tried to get at him. The man's eyes were bulging, and he grimaced. "My leg, my leg!" he shouted at one point. "——your leg!" one of his captors yelled back, and punched at him.

Just to the right, on the far end of the steam table near the door to the press room, a young woman, a Kennedy worker, held a man's head in her hands as he sobbed. "He's sick, he's sick," she whispered, apparently referring to the assailant. It made no difference to the weeping man.

Finally two medical attendants, dressed like police, came through the kitchen, rolling a low hospital stretcher. According to later Grand Jury testimony, they arrived at 12:23 A.M., six minutes after receiving the ambulance call, apparently from Minasian, the banquet captain. Many of those in the room—themselves in shock and observing the bedlam in a bizarre atmosphere of suspended time—thought it surely must have been longer than that. Jean Smith, the Senator's sister, was also at his side by now. The attendants moved immediately to Kennedy and took him by the shoulders and feet. "Gently," Mrs. Kennedy told them, "gently." They lifted him quickly, and Dick Tuck lifted his legs. "Oh, no, no," Kennedy said, in pain, "don't." Then, after he was strapped onto the stretcher, he lost consciousness. A kitchen boy took a white towel and began to mop up the pool of blood that had formed under the Senator's head. A piece of ice, apparently used as a coagulant, lay there, a sickening pink color.

The stretcher was wheeled out through the kitchen to a service elevator, the attendants bustling through, ignoring Barry's pleas to handle the Senator gently. Accompanying the stretcher now were Mrs. Kennedy, Barry, Fred Dutton, Jean Smith, Blanche Whittaker, wife of the mountain climber, and Warren Rogers, Washington editor of *Look* magazine and a family friend. There were harsh words spoken in the elevator, and tempers flared as one of the attendants, Max Behrmann, loudly started giving orders on what was to be done. Ethel told him finally, "Please lower your voice," but he disregarded her. Mrs. Whittaker, frantic, finally slapped the man lightly on the mouth. He turned on her. "Don't do that again, lady," he warned, "or somebody will get a crushed head." It could have gotten worse, but Barry interceded as the elevator reached the ground floor. The stretcher was rolled down a ramp and it started to get away. "Somebody grab that thing!" Barry called, and Rogers reached out and slowed it down. "Let go!" the attendant said to him. But Rogers held on until the stretcher was off the ramp and at the ambulance.

The one attendant, Behrmann, began to bark orders again. "Only Mrs. Kennedy rides with him," he said. There were protests and Dutton finally resolved them by shoving the man aside and getting in. The attendants closed the doors before Barry could climb in too. Rogers grabbed him, hustled him into the front seat, got in and slammed the door. The driver, Robert Hulsman, pulled the ambulance away, with siren screaming and red warning light flashing.

In the back, there was more trouble with the attendant. Ethel and Dutton were leaning over Kennedy, who was having difficulty breathing. Dutton told the attendant, who grabbed a plastic oxygen mask and gruffly pulled it onto the Senator's head, right over the bullet wound. Ethel gasped and began to berate the attendant. Up front, the driver apologized for his colleague's manner.

In minutes, the ambulance arrived at the Central Receiving Hospital, a mile away, and Kennedy was quickly wheeled into an emergency room. On the way to the front door, Ethel spied a photographer with his camera poised and threw a brush block on him. Inside, she stayed in the emergency room, but the others left. A team of doctors and nurses was ready to administer to the Senator. Kennedy appeared to be dead or near death, but a doctor, Dr. V. Faustin Bazilauskas, soon realized he was alive. He slapped Kennedy's face, calling him "Bob" as he did it. Dr. Albert Holt jumped on a stool and began an emergency heart massage, pushing on Kennedy's chest. "At first he was pulseless, then his pulse came back and we began to hear a heartbeat and he began to breathe a little erratically," Dr. Bazilauskas told the Associated Press later. But Mrs. Kennedy was frightened. "She didn't believe that he was alive—sort of because she could see that he wasn't responding," he said. "So when we began to get a heartbeat, I put the stethoscope in her ears so she could listen, and she was tremendously relieved." A heart-lung machine was applied—to massage the heart while administering oxygen and giving artificial respiration.

Outside the hospital, a lay priest and friend of the Kennedys who had been at the hotel, Father James Mundell, was seeking admittance. Mrs. Kennedy had tried at the Ambassador to get him into the ambulance and then had signaled to him to follow to the hospital. When he arrived a policeman barred his entry, despite explanations, and finally struck him with his arm, knocking him back. Ethel, seeing the scuffle, rushed out and pushed the policeman, explaining that she wanted Father Mundell inside. The officer turned and gave her a forearm blow in the chest. Dutton came up and demanded that Father Mundell be admitted. Finally, the priest was brought inside, where he gave the Senator absolution. Shortly afterward, a priest from St. Basil Church of Los Angeles, Father Thomas Peacha, arrived and took over, administering Extreme Unction, the last rites of the Roman Catholic Church.

While doctors were struggling to keep Kennedy alive and his wife was being subjected to such incredible treatment, police finally had arrived at the Ambassador at around 12:30 A.M. to relieve Grier and the others of the suspect. Several Los Angeles police, after handcuffing the man, carried him out into the lobby, then hustled him down the circular stairway that winds to the lower concourse and out to the taxi entrance. Unruh was sticking with them. Many of the stunned bystanders were not aware at the time that the man in tow was the suspect. In fact, unbelievably, one reporter overheard one of the policemen taking him out ask another: "Who did he shoot?"

In the police car, Unruh repeated again, "We don't want another Oswald." One of the arresting officers, Arthur Placencia, told the Grand Jury later that at his partner's suggestion he pulled out his officer's field notebook and read the suspect his legal rights. " 'Do you understand your rights?' (I asked him). And he looked at me and I asked him again, and then he mumbled something. So I told him, 'Look, I will give them to you again.' So I read them off again. 'Do you understand your rights?' Then he looked at me and said, 'Yes.' And then I said, 'Do you wish to remain silent?' And he just didn't say anything. Then he looked at me again and he said, 'Yes.' Then I said, 'Do you wish an attorney present?' Then at that time he just said, 'Yes.' So after that I didn't ask him anymore. And then after that I asked Jesse Unruh, I said, 'By the way, who did he shoot?' And he goes, 'Bobby Kennedy.' And I said, 'Oh.' "

Back in the kitchen, Grier put his head down on the steam table where he had pinned the man, and sobbed like a baby, soft and quiet. Finally the hulking giant of the Rams' "Fearsome Foursome" stood, tears running down his black face, and let friends lead him into another kitchen room behind the corridor.

There he sat next to Jim Wilson, a CBS television cameraman—a tough, muscular, hard-bitten veteran of the news business with the face of a wrestler—and each, oblivious of the other, sobbed, head in hands.

Wilson, who had traveled thousands of miles with Kennedy and had taken thousands of feet of close-up film of him, was right behind the Senator when the bullets hit. Horror-stricken, his professionalism as a cameraman put him into a kind of automatic control, and although he froze in his tracks and turned his head away, he kept his finger pressed on the trigger of his camera until all footage was expended. Wilson's electrician, Bill Lewis, reached up at times and kept the camera focused on the scene. When there was no more film, Wilson hurled the expensive TV equipment down, knelt at Kennedy's side and pounded the concrete floor in anguish. After a minute or so, he jumped up and began pushing people back. One of the most dramatic scenes filmed that night showed him rising from where the Senator lay and plunging, arms waving, into the crowd of onlookers. Now, about twenty minutes after the shooting, he had subsided into quiet sobbing.

In the stunned ballroom, the mood had plummeted with the barrage of shots from gaiety to instant agony. Steve Smith coolly walked to the microphone and asked the crowd to disperse. Shrieks filled the room; girls fainted; men wept. The scene was immediately transmitted to the nation by television. It was seen in New York and Chicago and Dallas and Denver. And in the Beverly Hills Hotel not far away it was heard by David Kennedy, age twelve, playing with a monkey somebody at Malibu had given to the children earlier in the day. Bob Galland, a young college student from Leonia, New Jersey, hired as a companion to the children, walked over to the television set, turned it off, called to the baby-sitter and took David for a long walk. As soon as they could get there, Teddy White and Joan and Tom Braden, California friends, comforted the children. Nicole Salinger, Pierre's wife, arrived later and spent the night.

Ted Kennedy, who had gone to San Francisco to handle the election-night operation for the northern California headquarters, had just returned from a victory rally and was in his room at the Fairmount watching television. Dave Burke was with him. There was a commotion on TV, and they weren't sure at first if they had the Los Angeles scene or not. Then it was clear it was Los Angeles, and that Robert Kennedy had been shot. Ted just stood there as the word came. Then, to Burke, he said simply, "We better get down there." Burke called a Congressman and a jet fighter was readied at nearby Hamilton Air Force Base. The younger brother climbed aboard with Burke, Seigenthaler and Bob Fitzgerald, a Kennedy cousin and campaign aide. The pilot whisked them south to Los Angeles, where a helicopter awaited to take Edward Kennedy to the hospital.

Upstairs in the Kennedy suite at the Ambassador, staff and friends watched the television sets in horror. Goodwin, on the phone in Dutton's room, heard a girl run down the hall saying. "Someone's been hurt!" Then he heard Steve Smith's voice on television saying help was needed. "It was like seeing an old movie over again," he recalled later. He, Walinsky and others drove at once to the Central Receiving Hospital.

Some were not watching television. Back in New York, Lowenstein sat with his wife, Jenny, talking and awaiting the call from Kennedy. "I don't know why I love Robert Kennedy so much," he told her, "but I guess I love him as much as anybody in public life since Mrs. Roosevelt." Just then the phone rang. "He certainly chose a good line to come in on," Jenny Lowenstein quipped. Her husband picked up the phone, expecting to hear Kennedy's voice. Instead, it was a friend from the *Wall Street Journal,* Skip Martin. "Kennedy's been shot," he said.

In the Ambassador lobby in Los Angeles, the great bewildered crowds stood in shock, women crying and slumped against the walls and columns, men milling around aimlessly. One woman put her head down on the American Airlines counter and wailed. "My God, my God, what kind of country is this?" she asked herself audibly. It was a question spoken and unspoken everywhere in the hotel, in the city, nation and around the world on this night and in subsequent days of grief and mourning.

At one point, a television personality, one of those hundreds whose face is familiar but whose name is not, came into the lobby proclaiming, "He's okay, he's okay," and some applause broke out. But the shock, and the disbelief, would not vanish. At the edge of a large, ornate pool of water in the lobby, just to the right of the cashier's window, a woman balanced herself and dangled a set of rosary beads over her head.

"Kneel down and pray, kneel down and pray, say your rosary," she implored the crowd. Immediately, twenty or more persons knelt around the pool. One man holding a drink and a cigarette placed both down on the red and black carpet and knelt. Another man, wild with frustration, picked up a chair and hurled it into the fountain. Others seized him and calmed him down.

Outside the hotel, near the parking lot, a fight broke out when somebody said, "Men are dying every day in Vietnam." The combatants were separated by the crowd.

Rumors swept the hotel. Others had been shot: Jesse Unruh and Steve Smith were said to be among them. This was not so, but there were five other victims. Paul Schrade, the forty-three-year-old UAW official Kennedy minutes earlier had thanked from the speaker's platform for his help in the California victory, lay in the kitchen corridor, about three feet behind and to the right of where Kennedy had fallen. He too had blood rushing from his head and appeared to be seriously wounded. Somebody lifted his head and placed a Kennedy campaign straw hat under it. The blood soon turned the hat a grisly red. He was taken out shortly after the Senator was removed. Schrade wore an "Embassy Room Press" badge, leading some to believe at first that he was a reporter. But nearly everybody, it seemed, had a press badge that night. There were security guards at all regular entrances to the ballroom, but badges to gain entry were a dime a dozen.

In the little anteroom just behind that area, where the two women had been sitting so nonchalantly as Kennedy came through, the others who were shot had fallen or were staggering about, holding their wounds. William Weisel, thirty, associate director of the Washington bureau of ABC News, was hit in the left side near the abdomen. Ira Goldstein, nineteen, of Continental News Service at Sherman Oaks, California, was shot in the left thigh. Elizabeth Evans, forty-three, of Saugus, California, was grazed in the right forehead by a bullet that lodged in her outer skull at the hairline. She did not even fall; she wandered around, holding a handkerchief to the wound. A seventeen-year-old-boy, Irwin Stroll of Los Angeles, was shot in the left shin but not seriously hurt. All five were moved out quickly after Kennedy had been attended to. All survived.

Later, before the Grand Jury, young Stroll told how he had "got in front of Mrs. Kennedy by accident" going into the kitchen. "We went in the door and all of a sudden the procession stopped and it was like firecrackers, just pop-

pop-pop all over the place—and smoke—and well, I did this, I just pushed Mrs. Kennedy, just a reaction. I turned around and saw Mrs. Kennedy on the floor with Roosevelt Grier covering her. It was like a swarm of people pushing me out of the room. I went out the podium door. I said, 'God, I'm shot!' "

Goldstein told of being hit in the left thigh and staggering to a chair: "(I asked) 'How is Senator Kennedy? What happened to him?' And this woman walked by, and she said to me, 'How dare you talk about my husband that way?' and she slapped me across the face. And I said, 'I am sorry, lady, but I was shot too. I'd like to know how the Senator is.' And she said, 'Oh, I am sorry, honey,' and kissed me. This was Mrs. Ethel Kennedy. At that time she was not in tears. She was a little hysterical, though, but she wasn't crying."

At the Beverly Hilton Hotel far down Wilshire Boulevard, Senator McCarthy got the news as he sat in his seventh-floor suite with Blair Clark and Thomas Finney, two top campaign aides, drafting a congratulatory telegram to Kennedy. Mary McGrory of the Washington *Star*, a McCarthy confidant, also was there. The draft referred to Kennedy's "splendid" victory, based on the early—and overstated—predictions of Kennedy's margin, and McCarthy had been balking at the word.

"I think that we could say 'fine' instead of 'splendid,' " Miss McGrory reported him as having just said, "because I don't think the percentage will go that high." (He was right; the final figures gave Kennedy a margin of only 4.5 points, 46.3 percent to McCarthy's 41.8 percent, with the rest for the pro-Administration slate of State Attorney General Lynch.)

Suddenly at the door was Schoumacher, the CBS man with the McCarthy campaign. "Senator Kennedy has been shot," he told them.

"You're kidding," somebody said.

"I'm not," he answered. "I'll go back and get more."

Abigail McCarthy, the Senator's wife, and their daughters, Ellen and Mary, came in from the next room. McCarthy, in a corner chair, put his hands over his eyes, then looked up, in disbelief. "Maybe," he said, "we should do it in a different way. Maybe we should have the English system of having the Cabinet choose the President. There must be some other way." Outside the room, police set up a guard in the corridor.

Downstairs in the McCarthy headquarters ballroom, the Senator's ever-optimistic young workers were singing "When the Saints Come Marching In," when the news was announced. A pall shrouded the scene. One McCarthy aide was heard to say: "If he lives, I'll work for him."

But the chances that Kennedy would live already were grim. One bullet had entered the mastoid area behind his right ear, on a path to his brain. After the emergency treatment at Central Receiving Hospital, it was decided to move him to the Hospital of the Good Samaritan several blocks away for surgery.

At about 12:50 A.M., Dutton came out and told newsmen the Senator was going to be transferred. Jean Smith walked out and got into a car. As she did, two Negro women ran over and said to her: "God bless you, sweetheart, we'll pray for you! We love you, we love you!" And then they literally began to roll in the grass, moaning and crying.

Within ten minutes, Kennedy was brought out on a stretcher, drains in both nostrils, an oxygen mask over his face, bottles of plasma and dextrose being administered en route. Mrs. Kennedy walked directly alongside his head as he was wheeled to the ambulance. She got in. Barry and Dave Hackett, another long-time friend, took the front-seat places next to the driver. They sped off to Good Samaritan, a nine-story boxlike structure of gray stone with a large white cross and host on the roof, spotlighted. The Senator was taken to a fifth-floor intensive care unit, then moved to the ninth floor for surgery. Mrs. Kennedy went to a nearby room and was persuaded by Father Mundell and others to lie down for a rest. Other members of the family, including Jean and Steve Smith and Pat Lawford, also were in the hospital.

Outside, the vigil of newsmen and neighbors already had started. Television crews moved in quickly and set up their equipment. In a doctors' parking lot up a driveway directly across from the entrance, mobile trucks pulled in and technicians ran thick cables down over the sidewalk to the street. Police kept everyone on the opposite side from the main entrance, and a Los Angeles local TV truck focused a large floodlight on it. All approaches to the streets on both sides of the hospital were quickly blocked off. Red flares on the ground cast an eerie light on the faces of Los Angelenos who came to the hospital and stood wanly behind the barriers, waiting for some word or sign. Throughout the building, lights blazed. Hospital workers and patients looked out their windows at the crowd, which had become so noisy that Dutton finally came out to plead for quiet.

More than an hour passed before Mankiewicz emerged at 2:22 A.M. He described Kennedy's condition as "very critical," and said surgery would begin in five or ten minutes for an operation of "forty-five minutes or an hour."

Actually, the operation did not start until 3:12 A.M., and lasted not forty-five minutes, but three hours and forty minutes. A team of three neurosurgeons and three chest surgeons from the medical faculties of the University of California and UCLA performed the operation, under the direction of Dr. Maxwell M. Andler, a UCLA neurosurgeon. He was assisted by Drs. Henry Cuneo and Nat Downes Reid, the other neurosurgeons; and Drs. Burt Meyer, Paul Ironside and Hubert Humble, the chest surgeons. Dr. James Poppen of the Lahey Clinic in Boston was flown in on a plane made available by Vice President Humphrey, but he arrived after the surgery was completed.

There were three wounds. One bullet had passed the mastoid bone and lodged in the midline of the Senator's brain. Another lodged in the back of his neck and a third grazed his forehead. The first was the critical one.

During the long wait, a steady stream of familiar faces in the Kennedy campaign moved in and out of the hospital—Dick Tuck, Unruh, John Glenn, Andy Williams, Goodwin, Rafer Johnson, finally Rosey Grier, able at last to curb his grief outwardly. Only the immediate family was near the Senator, in surgery down the hall. From time to time, one or another of the doctors would provide a briefing, through Steve Smith, Salinger or directly. Salinger took the lead in insisting that the doctors get together on what the situation was, and then have that one consensus made public through Mankiewicz. And outside, darkness drained away to early dawn as newsmen occupied themselves comparing what they had seen and heard, reflecting on Kennedy's own sense of fatalism, recalling similar moments in Dallas in 1963, anticipating the worst as the time dragged on.

Back at the Ambassador, hangers-on still were wandering around the Embassy ballroom. Empty glasses and paper cups stood on tables in silent recollection of the moments of elation a few hours before that had turned suddenly into stark tragedy. A man walked over to a large campaign poster of Kennedy, peeled it gently and with affection from the wall, rolled it carefully and walked away with it.

In the kitchen corridor where Kennedy had been hit, the path he took was strewn with cigarette butts up to the point where he and Schrade had fallen. At that point the floor had been swept clean and two large white X's in chalk showed where they had been lying—more antiseptic markers than the pools of blood that had been soaked up and scrubbed away in the intervening time, with what seemed somehow like indecent haste. Overhead, several dry-wall panels in the ceiling had been removed in the search for bullets, and police now stood guard over the nothingness that was left. Chuck Quinn of NBC waited there forlornly to give a television report from that now-placid scene, and he had it all to himself.

On a wall beyond the point where Kennedy had been shot, a hand-printed sign said THE ONCE AND FUTURE KING, a reference to the book from which the musical *Camelot* was taken. It was not known whether it had been placed there before or after Robert Kennedy was shot.

Across the country, Americans in the East were waking to the shocking and bizarre news. President Johnson had been called by Walt Rostow, his national security assistant, at 12:31 A.M. Pacific time (3:31 A.M. Washington time), about the time Kennedy was being taken from the Ambassador, and he began to confer by phone with Secretary of Defense Clifford, Attorney General Ramsey Clark, Secret Service Director James J. Rowley, FBI Director J. Edgar Hoover and members of Congress. At 3:45 A.M. (6:45 A.M. in the East), he issued a statement. "There are no words equal to the horror of this tragedy," the President said. "Our thoughts and our prayers are with Senator Kennedy, his family and the other victims. All America prays for his recovery. We also pray that divisiveness and violence be driven from the hearts of men everywhere." Then he dispatched personal messages to Ethel Kennedy and the Senator's parents. Mr. and Mrs. Joseph P. Kennedy, and called Ted Kennedy and Steve Smith at the hospital.

At Hyannisport, the phone rang and Ann Gargan, Ambassador Kennedy's niece and companion, answered. Ted Kennedy was on the other end, and he told her and his father. Miss Gargan informed Mrs. Rose Kennedy, the Senator's mother, before she went to Mass. The incident later gave a newsman an opportunity to demonstrate his insensitivity, and he came through with flying colors, asking Miss Gargan how the elder Kennedys had taken the news.

"How would you take it if you were told another son had been shot in the head?" she snapped back. Then her voice broke and she added: "It's terrible. We are praying."

At Hickory Hill, friends began to arrive to be with the two Kennedy children at home, Robert Jr., fourteen, and Douglas, a year old. The first came at about an hour after the shooting, or sometime after 4 A.M. Eastern time. The only light in the big colonial house at the time was in a top-floor bedroom, and a maid admitted them.

About forty-five minutes later Melody Miller, a pretty strawberry blond receptionist in the Senator's Capitol Hill office, came to the door, was identified by a cook, and admitted. Then came Laverne Duffy, assistant counsel of the Senate Rackets Investigating Subcommittee, who told police on duty that he often stayed with the Kennedy children when their parents were out of town. He too was admitted after the cook saw him. Others arrived afterward in a steady stream, as dawn approached. At about 8 A.M., Robert Jr. was seen walking in the sprawling backyard, talking quietly with two men. He flew to Los Angeles later Wednesday, along with his sister Kathleen and brother Joseph. Mrs. Rowland Evans, wife of the columnist, escorted them.

Around the world, friends and enemies of the United States awoke with shock and sympathy at the news. In Europe, even before the assailant had been identified, talk of a "plot" spread everywhere. The various theories that John F. Kennedy had been assassinated as part of a political plot, foreign or domestic, had gained early and wide currency in Europe, and it was starting all over again. In Vatican City, Pope Paul VI appeared before a large crowd in St. Peter's basilica and said in English that he was praying "for the life and health of this young man who was offering himself to the public service in his country." Many in the throng, not having heard the news, gasped; so did many more when he repeated it in Italian. And in London, Prince Stanislaus Radziwill received a transatlantic call and immediately phoned his sister-in-law, Jacqueline Kennedy, in New York. She had been sleeping. "You heard he won in California?" Jackie asked. "Yes," Radziwill answered, "but how is he?" "I just told you," Jackie insisted. "He won in California." Obviously, she didn't know. Radziwill broke the news and then rushed with his wife for a plane to join the stunned widow of the late President.

But now, in this era of instantaneous communications and transoceanic television by satellite, none outside the Hospital of the Good Samaritan—not those standing wearily across the street, not those halfway around the globe listening to their radios and watching television—knew any more about Robert Kennedy's battle for life.

At about 4:45 A.M., Pacific time, Mankiewicz came out, grim and puffing tightly on a cigarette. He approached the crowd of newsmen, took a final puff and threw the cigarette away. He picked up a police microphone; it barked, so he climbed atop the roof of a police car. "The doctors now say that the surgery will take another hour and perhaps two." It sounded ominous. Mankiewicz added that the Senator's "life signs"—heartbeat, respiration, blood pressure—remained "good," but he would say no more. He hopped down and went back inside.

It was 7:20 A.M. in California—10:20 A.M. in the East—when he returned with a report. All but one fragment of the critical bullet had been removed from Kennedy's brain. The bullet in his neck had been left there because it was "not regarded as a major problem" by the doctors. The Senator's life signs were about the same, except that now he was breathing on his own. But, Mankiewicz concluded, "there may have been an impairment of the blood supply to the midbrain, which the doctors explained as governing certain of the vital signs—heart, eye track, level of consciousness—although not directly the thinking process." The words focused on what had crept into everyone's thoughts: even if Kennedy should live, would he ever be able to function again? There was no answer to that. The next twelve to thirty-six hours would be "a very critical period," Mankiewicz said. No additional surgery was contemplated.

Inside, back on the fifth floor, near the intensive care unit, the same thoughts were playing on the minds of those who had been closest to Robert Kennedy through his forty-two years, or in the later political years. Already, by the grimness that often communicates bad news in hospital corridors, these vigil-keepers could sense that the outlook was worse than indicated by the antiseptic words passed on by the doctors. One early report had said that there had been no brain damage, and some clung to that, with hope. But there were nurses, and other doctors, and police and Secret Service men, and talk drifted up and down the corridor with a depressing air of inevitability.

Now the real vigil set in. Dick Drayne and Hugh McDonald came out and helped set up a press room on the ground floor of the hospital, then had to move it across the street, to the gymnasium-auditorium of a hospital dormitory for residents, because police were edgy about security within the hospital.

The same edginess prevailed at police headquarters downtown, where word on the identity of the suspect was awaited. Several hours after his arrest, it came from a most unusual source—the mayor of the city, Samuel W. Yorty. The mayor, a Democrat but a longtime bitter political foe of Kennedy's, held a press conference at which he identified the man as Sirhan Bishara Sirhan, twenty-four, a Jordanian born in Jerusalem who had lived in the Los Angeles area since 1957. The man was quickly arraigned at 7:40 A.M., Pacific time, and charged with six counts of assault with intent to murder. Bail was set at $250,000. Los Angeles Police Chief Thomas Reddin said the suspect's identity had been traced through the revolver and fingerprints taken when he had applied earlier for a job as an exercise boy at Hollywood Park racetrack. Two brothers confirmed his identity, Reddin said. First inquiries were said to have revealed no motive, but a newspaper article critical of Kennedy, and four $100 bills, were found in his pockets. Sirhan apparently was not talking.

But Yorty was. Two notebooks had been found in the suspect's Pasadena home, he reported later in a television interview, that had "a direct reference to the necessity to assassinate Senator Kennedy before June 5, 1968"—the first anniversary of 1967's six-day Arab-Israeli war. The notebooks, Yorty said, also contained "generally pro-Communist writings" and pro-Arab, anti-Israel, anti-Kennedy references.

Yorty's off-the-cuff revelations were only the first indiscretions by him that infuriated the Kennedy family in the hours and days after the shooting. Chief Reddin indicated shortly afterward that he thought the mayor should have refrained from discussing the notebooks because his remarks might prejudice Sirhan's trial. As for the man's security, Los Angeles police were doing their utmost to make sure that the nightmare of November, 1963, in Dallas did not repeat itself in any further aspects in their city. They kept the suspect under heavy guard and out of public view or access.

At the hospital, although there was no more official news, the atmosphere was increasingly somber and pessimistic. Some other Kennedy aides thought Mankiewicz' reports were putting the best possible light on the Senator's condition, and that it seemed hopeless. One of the doctors who assisted in the surgery, Dr. Cuneo, apparently had called a colleague in New York, Dr. Lawrence Pool, chief of neurosurgery at the Columbia Presbyterian Medical Center; Dr. Pool, interviewed on television afterward, quoted Dr. Cuneo as saying it was unlikely Kennedy "will be able to recover fully" and that his condition "makes it doubtful that he could live."

During the morning, Senator McCarthy came to the hospital to pay his respects to the Kennedy family. He talked to Salinger and his old ally, Goodwin, then left for Washington, calling a halt to his own campaign.

All through the warm, bright day of official silence, crowds milled quietly outside the hospital. Although it had been less than eighteen hours since the shooting, in the late afternoon orange bumper stickers with black letters reading PRAY FOR BOBBY began to appear in the hands of Negro women and on the backs of young boys in the hospital area and around the city.

In the press room, hard-bitten reporters, taking a cold reading of the situation, sat at their typewriters and began to write the last rites of their own profession, the obituary. Many others knew they should set about the task but could not bring themselves to it. Over Kennedy's fifteen years in public life, they had come to know him with understanding, appreciation and warmth; he was a political figure, a candidate for the Presidency, but also for many of them, particularly those in the Washington press corps, he was a contemporary coming of age in comprehension of the nation's political experiences they had shared with him. No other group had spent more time with Kennedy during the campaign than the press. While key aides labored back in headquarters or around the country, the newsmen plodded along day after frenetic day at his heels. One day on the campaign plane, Tuck had clipped the Spanish words *Viajando Juntos* from several small airlines cards used to reserve seats, and passed them out to reporters to paste on their "Kennedy Press" badges. The words meant "traveling together," and that's how it was always.

One of the newsmen had talked of the relationship between the reporter of integrity and Robert Kennedy in a conversation on the Kennedy press bus only two days before—the last campaign day in California. "I hope I don't have to cover Kennedy too much," he said, "because he gets to me—you have to guard against liking him too much." It was an admission some reporters would never have made, either because they didn't like him, or more frequently, they liked him too much and showed it, by becoming de facto members of the entourage. Within the press corps, there were those known to be "soft on Bobby," but it was easy to be that way, especially in the later years and months. And now particularly, it was difficult to keep Robert Kennedy in perspective and to write about him in the past tense—when there was at least a thread of hope left.

In Los Angeles and elsewhere that afternoon, the search was on to determine the background of the suspect. His distraught mother, his neighbors in Pasadena, his former classmates and employers, even his father, separated from the family and living in Et Taiyiba, Israel-occupied Jordan, all were sought out. The consensus collected was of a young and intense but friendly boy, resistant to authority, with a bitterness toward Israel and those who supported it, stemming from his early days in Jerusalem.

Reactions to the shootings varied. In Washington, government workers picketed the National Rifle Association headquarters, vanguard of the gun lobby. In Roseburg, Oregon, S. J. Schoon, national director of the National Association to Protect Our Right to Keep and Bear Arms, the man who had engaged Kennedy in debate on gun control just the week before, said: "I consider it a very tragic shooting. I hope that the tragedy does not spur an even greater tragedy, that people will return to a respect for law and order, and that through emotionalism new [gun control] laws are not speedily passed."

It was 5:30 P.M., Pacific time, before Mankiewicz had any more news. He walked out of the hospital, crossed the street and went directly to the microphones set up in front of the stage in the makeshift press room. He was grim, and those who knew him braced themselves.

"The team of physicians attending Senator Robert Kennedy is concerned over his continuing failure to show improvement during the postoperative period," he said. "Now, as of 5 P.M., Senator Kennedy's condition is still described as extremely critical as to life. There will be no further regular bulletins until early tomorrow morning." And he left.

The key phrase was "as to life." It meant that no longer was the primary question whether Kennedy could recover without impairment of his faculties, but whether he could survive the surgery at all. The doctors had been counting on his excellent physical condition prior to the shooting to help him pull through, but he was not bouncing back.

Shortly after Mankiewicz left, word came from Salinger for the press to stand by for an announcement. Mankiewicz had said no "regular" bulletins until morning; the worst anticipations seemed about to be realized. But for seven more hours nothing came. In the early evening, Mrs. Jacqueline Kennedy's arrival caused a rush of curiosity-seekers toward her car outside the hospital. As the nation waited, the television networks continued to show reruns of the bizarre scene in the Ambassador kitchen and, at one interval, a special appearance by President Johnson announcing the appointment of a commission to study violence—with Senator Roman Hruska of Nebraska, champion of the gun lobby, as a member. Inside the hospital, Kennedy intimates watched from their fifth-floor post. One of them later compared Johnson's strident voice, coming over the television in the waiting room, to "a fingernail scraping across a blackboard."

There, inside Good Samaritan, they knew from early evening that it could not be much longer. Machines recording the Senator's brain waves showed a flat even line by then, indicating the brain had ceased to function, while other devices kept his heart pumping and his blood circulating. At one point, the closest of his campaign aides were permitted into the room to be with him a last time. He was unconscious as they said their private, silent good-byes.

Finally, twenty-five hours and forty-seven minutes after it all had started in the Ambassador kitchen corridor, Mankiewicz entered the press room for the last time, at 2 A.M. Weary, his voice choking but under control, the indomitable press secretary said simply: "I have a short announcement to read which I will read at this time. Senator Robert Francis Kennedy died at 1:44 A.M. today, June 6, 1968. With Senator Kennedy at the time of his death were his wife, Ethel; his sisters, Mrs. Stephen Smith and Patricia Lawford; brother-in-law Stephen Smith; and Mrs. John F. Kennedy. He was forty-two years old." Mankiewicz later added the name of Senator Edward Kennedy.

Everyone had known it was coming, but faces and hearts sank nonetheless. Somebody asked what the specific cause of the Senator's death was, and the press secretary, with a look of numb disbelief, replied: "the gunfire attack . . . the bullet that went into the head near the right ear."

Outside, under the floodlights across from the entrance to the hospital, women in the crowd wept. At the White House, President Johnson said: "This is a time of tragedy and loss. During his life, he knew far more than his share of personal tragedy. Yet he never abandoned his faith in America." The President proclaimed the next Sunday as a national day of mourning, and all flags on all public buildings were ordered lowered to half staff.

The family remained at the hospital the rest of the night, while friends returned to the Ambassador and packed for them. Later on, a full six-hour autopsy was performed on the Senator's body, to meet California legal requirements and to make sure there would be no repetition of the controversy that occurred after his brother's death in Dallas. And at the Los Angeles County jail, the $250,000 bond on Sirhan Bishara Sirhan was revoked and plans were made to seek a grand jury indictment of murder against him.

Now all that was left was the ritual of placing another Kennedy to rest. A Presidential plane would be dispatched to Los Angeles by the White House to take the body and the family to New York, Robert Kennedy's adopted state. Another Presidential jet, to another city, in another time; but it was all so bizarrely the same. The next morning, on the commercial flight from Los Angeles to New York, some reporters, still wearing their Kennedy campaign badges that said *Viajando Juntos*, decided to watch the in-flight movie as a temporary distraction as they rode east to meet the Kennedy plane on arrival at LaGuardia Airport. It was a spy drama—routine American entertainment fare for all the family. In an opening scene, the head of a very lifelike dummy was blown off by gunfire.

CHAPTER NINE

The Last Farewells

THE LONG TRIP HOME

AS SOON AS it was over, and the word had been transmitted to a nation waiting in dismay and to an uncomprehending world beyond, there was, as there had been the last time, the sorrowful business of bringing the assassinated leader home. Now, for the first time, that same long ugly adjective could be used. In 1963, in Dallas, the nation's continuity was involved; this time it was just its grief. He was a Senator and a candidate for the Presidency, not the President, as the other had been. But it was a national tragedy of incalculable dimensions because they were brothers, and because the one had tried to carry on in the national life what the other had not been permitted to do. And in doing so, he had touched the past pain of America, and for many, assuaged guilts and rekindled hopes. Messages of sorrow and condolence poured in. Senator McCarthy said: "Words can do little to ease the anguish of the family of Senator Robert F. Kennedy or the grief of all those who loved him. Let us seek to comfort them by our quiet mourning, our rejection of violence and reprisal, and by offering renewed dedication to the cause of peace and reconciliation which Robert Kennedy served." McCarthy then ordered his campaign workers to suspend all activity until further notice.

Now as in that earlier time, the strong ones moved in to help the most bereaved, to make all the businesslike and ritualistic decisions that custom and the wishes of the lifelong devout require. The last time it had been Robert Kennedy, the second brother, who had been the strong arm of the first's bereaved widow; now it was the third brother who moved to the side of the second's wife.

And as always, in life or death, the sureness of the closest brother-in-law, Steve Smith, gave direction and order to the others who stayed together to finish this one last assignment.

At a nurses' counter down the fifth-floor corridor, and in two small adjacent rooms, the campaign team labored as instructions were moved to them from where Ethel and Ted Kennedy, Steve and the sisters gathered. The basic decisions were submitted to Ethel for approval; some originated with her, like the choice of St. Patrick's Cathedral for the Solemn Requiem Mass and of Andy Williams to sing there, and the burial in Arlington.

Two working groups were called together, one in Washington and the other in New York, to make local arrangements and to begin the massive sorting-out job of invitations to the church and to the burial, and the even more staggering logistical tasks in transporting the thousands who knew Robert Kennedy as a friend or as a political leader.

In Washington, old Justice Department aides like Undersecretary of State Nicholas Katzenbach, John Nolan and Louis Oberdorfer pitched in; in New York, vanden Heuvel and others rounded up the dependables. Oberdorfer undertook the task of determining the invitation list, with help from vanden Heuvel. Salinger, from the hospital, called Vice President Humphrey's office and arranged to have one of the Air Force Presidential planes flown to Los Angeles to take the body and the family and closest friends to New York. It was the same model as the Air Force One that brought John Kennedy home from Dallas, but not the identical plane.

The decision to move the Senator's body and the accompanying mourners from New York to Washington by train evolved more out of logistical realities than from any intent to duplicate the lonesome train of Lincoln. The Washington group first was thinking of a plane from New York, but the thousands of persons involved made it impractical.

The family was not immediately certain whether the Senator could be buried at Arlington, though he was a Navy veteran of World War II, or whether he could be placed near the grave of his late brother, the President. Congress had imposed new limitations on use of the most famous of all the national cemeteries, because of the high demand for space. To resolve both issues, Steve Smith phoned former Secretary of Defense Robert S. McNamara in West Germany, where he was en route to Indonesia on business as the new director of the World Bank. McNamara received the call about four hours after Kennedy's death, was crushed by the news and took a plane back home immediately. Before leaving he called the Pentagon, where Assistant Defense Secretary Alfred Fitt, who oversaw matters concerning the Kennedy gravesite, already was at work. After checking with Deputy Secretary of Defense Paul Nitze and Secretary Clark Clifford to make sure Kennedy could be buried in Arlington, Fitt went to the Kennedy gravesite and began at once to look for the best location.

So efficient was the far-flung Kennedy team in this last massive planning effort that Salinger, walking into the press room at the hospital less than two hours after Mankiewicz' announcement of Kennedy's death, was able to provide a detailed outline of what would be done, where and when.

While this necessary, unseen work went forward, the drama of a nightmare reenacted continued to unfold. In the latest assassination city, a public official, this time Mayor Yorty, made more public statements about the suspect. In spite of criticism from State Attorney General Lynch that Yorty's revelations of the previous day might impair a fair trial, the mayor held another televised press conference at the Police Administration Building and added more observations about the "Communist sympathies" of the man in custody.

And already the conspiracy rumors were in the making. At the county sheriff's office, a nineteen-year-old model who gave her name as Kathy Fulmer walked in and said she possibly was being sought by authorities. The Los Angeles *Times* had printed reports of a mysterious "girl in a polka-dot dress" running from the Ambassador shortly after the assault, shouting "We shot him!" Young Vincent DiPierro, later Grand Jury testimony also revealed, had seen "a very good-looking girl" standing next to the suspect in the kitchen corridor wearing "a white dress [with] either black or dark purple polka-dots on it." The suspect had "either talked to her or flirted with her," DiPierro said, "because she smiled. Together, they were both smiling; as he got down he was smiling. In fact, the minute the first two shots were fired, he still had a very sick-looking smile on his face. That's one thing—I can never forget that." The sheriff's office questioned Miss Fulmer and released her—to news photographers who duly snapped her for the awaiting world to see.

The focus of the nation, however, remained on the Kennedy family, again embarking on a public journey of tragedy. When the Senator's body was removed from Good Samaritan around midday, spectators stood outside the hospital, along the route and at Los Angeles Airport. Many wept and tossed flowers as the hearse drove slowly by, his widow and his brother riding in the front seat. Mankiewicz, arriving at the airport in advance of the party, saw Mayor Yorty waiting near the Presidential jet. The haggard press secretary informed him his presence would be "upsetting" to Mrs. Kennedy, whereupon Yorty went inside the terminal.

At planeside, the casket was placed aboard, in the front compartment, while passengers entered by another door. Jacqueline Kennedy, assured this was not the identical plane that brought her husband home from Dallas, boarded and sat with Stash Radziwill. At 1:28 P.M., Pacific time, as the Presidential jet took off and moved with silent swiftness across the great, shocked country bearing the body of a second assassinated Kennedy, many of the same sensitive individuals who were on the earlier flight from Dallas sat gripped in numbing disbelief. A cold buffet and drinks were served, but many aboard, exhausted, slept through. Ted Kennedy, and often one of the close friends, sat up front with the casket at all times. At one point, the surviving brother called to his side each of the newsmen aboard to talk of the past and the future; he stressed his increased family responsibility, both personal and political. It seemed clear even then he had dismissed thoughts of any active role for himself in 1968. Ethel, when she was not at the casket, walked up and down the aisle, talking to the passengers. She solicitously inquired about where her friends were going to sleep in New York. The Kennedy sisters and the three oldest children sat in the first-class compartment directly behind the coffin. The Justice aides set about organizing the honor guard for the night at St. Patrick's.

It was shortly before 9 o'clock at New York's LaGuardia Airport when the giant Presidential jet streaked over the field. For more than an hour in the warm spring evening, national and local notables had stood inside the gate at the Butler Marine Terminal, waiting, exchanging hushed conversation. Mrs. Edward Kennedy had been the first to be seen; she melted into the crowd of famous faces: United Nations Ambassador Arthur J. Goldberg, Governor and Mrs. Nelson A. Rockefeller, Mayor and Mrs. John V. Lindsay, Senator and Mrs. Jacob K. Javits of New York, Secretary of Interior Stewart Udall, Nick

Katzenbach and a host of Kennedy aides, some of whom had been in Los Angeles and had come east earlier: Walinsky, Edelman, Greenfield, O'Donnell, Sorensen, Schlesinger, Angie Novello. And in the incomparable pattern of New York politics, major and minor local hacks insinuated themselves onto the scene, making themselves as conspicuous as possible there in full view of the television cameras.

McNamara, having arrived at Kennedy Airport only shortly before, had been met by Fitt in a small Jetstar and whisked to LaGuardia. McNamara, who was at Andrews Air Force Base to meet the body of John Kennedy in 1963, had insisted he be on hand for Robert's last return. Fitt by now had detailed sketches and photographs of gravesite possibilities for his former boss to submit to Mrs. Kennedy.

A gray hearse was parked inside the gate and a phalanx of black limousines behind. A dolly covered in maroon velvet waited to take the coffin. The plane touched down in the waning moments of dusk and taxied over to the gate at 9:03 P.M. behind a yellow lead truck. The nose of the huge silver and blue bird swung around to the left of the crowd and parked in the very same position that earlier Presidential jet had assumed at Andrews on that night in November, 1963. This particular plane, identified by the number 86972 on the tail, also had been involved in the previous drama. It was taking Secretary of State Dean Rusk, Salinger and others to Japan when the news came, and it quickly returned home. Now, at LaGuardia, once again necks craned to view the faces of the anguished family and this nightmarish montage of history repeating itself. Floodlights, as they had done on that earlier night, pushed aside the growing darkness and robbed the homecoming of even the barest privacy. Not a tear, not a gesture, not a solitary misstep could escape the notice of crowd and camera.

The rear door, just below the huge number and the reversed American flag on the jet's high tail, was the first to open. Three airmen came out and then a stream of familiar figures, moving quickly down the stairway: Mankiewicz, Pierre and Nicole Salinger, Dick Tuck, Ed Guthman, Jerry Bruno, Jim Tolan, Rowland Evans, Joe Kraft, Sandy Vanocur, Rafer Johnson, others. As they descended, the front door opened and a hydraulic lift with a boxlike van—the kind of device used to put hot food aboard—was moved over. In 1963, they had brought the coffin out a rear hatch and had great difficulty making the descent, once or twice seeming to be on the verge of dropping the casket. But the planners, insensitive as it sounds, had experience now and a bit more time to make arrangements. Eunice and Sargent Shriver and Joan Kennedy were standing in the van as it moved to the door, where the coffin already had been placed, a maroon blanket over it.

Bill Barry, standing at the head of the casket, put his large strong hands around it and began to lift and guide it onto the awaiting vehicle. Ted Kennedy, standing just behind Barry, moved forward and also took a grip on the casket, and he and others helped Barry place it aboard. Ethel Kennedy, wearing a Navy blue coat and showing no signs of breaking, followed the casket; with her were the other sisters, Pat and Jean; sister-in-law Jacqueline, dressed in dark brown and looking ashen; the deceased's two oldest boys, Joseph and Robert Jr.

The vehicle's accordionlike elevator descended slowly, bringing the party to the ground, where a small white wooden staircase was laid against it. Barry and Ted Kennedy moved the coffin down carefully, with the Kennedy sons on the sides and Rafer Johnson and Jim Whittaker, who had climbed Mount Kennedy with Robert, at the rear. They lifted it gently on the dolly. Waiting were the Most Reverend Terence J. Cooke, Archbishop of New York—he who had greeted Robert Kennedy on the steps of St. Patrick's on the day the candidate launched his campaign—and other priests. Sprinkling holy water on the casket, the Archbishop recited the Catholic Ritual for the Burial Rites of an Adult:

"Oh Lord, we commend to you the soul of your servant Robert, that having departed from this world, he may live with you and by the grace of your merciful love, wash away any sins that in human frailty he may have committed in the conduct of his life, through Jesus Christ, our Lord."

The dolly was wheeled over to the rear of the hearse, and the coffin placed aboard. The widow and the surviving brother rode with it, and other family members and friends entered limousines behind. Joseph and Robert Jr. walked far to the rear and began to get into a car there, but were called forward. Jackie Kennedy, accompanied by Radziwill, moved down to the sixth or seventh car and got into the rear seat, on the side away from the plane. A moment or two later, McNamara came along on the opposite side. Jackie climbed out of her car and McNamara, spying her, rushed over, leaped up onto the hood of her car and down to her side, embracing her. There was anguish on her face, but not tears—a grim reminder of that similarly tragic night when her own husband was brought home for the last time, and she in utter shock with him.

Silently, the motorcade of twenty-seven vehicles moved out behind four motorcycle escorts and a police squad car, through heavy New York traffic, across the Triborough Bridge, onto the East River Drive, and crosstown to St. Patrick's at Fifth Avenue and East 51st Street. Small groups of people lined the streets all the way, and crowds packed the highway overpasses. New Yorkers waved silently from elevated railway stations, and going through Spanish Harlem on East River Drive, hundreds of Puerto Ricans ran up shouting "*Viva!*" and waving.

The night was hot and muggy, the kind of night that drives millions of New Yorkers to air-conditioned restaurants, movies, bars or other cool retreats. Many no doubt were finding refuge in such places, but an estimated 7,000 others already had flocked to the streets on both sides of the huge cathedral, standing quietly behind police barricades, enduring the oppressive heat. Fifth Avenue was closed to traffic for several blocks on either side of the church. The hearse and the motorcade pulled up and the family and close friends followed the casket up the steps and into the church.

Already waiting there was Mrs. Rose Kennedy, the deceased's mother, in black dress, shoes and veil. Virtually unnoticed, she had arrived in a police car, had ducked under a barrier and gone inside. The coffin was placed in the transept and a brief prayer service was held by Archbishop Cooke. Then the closest associates of Robert Kennedy formed an honor guard that was to keep vigil over him all night, until St. Patrick's doors were swung open for the public. Bruno and Dutton took the first turn along with some self-starters from the New York political crowd.

The family left through the front door—Ethel Kennedy with Ted Kennedy and her eldest son, Joseph, at her side; Robert Jr. escorted his grandmother. Then Jacqueline Kennedy, flanked by Radziwill and Burke Marshall, appeared on the top step. She stood there for an excruciatingly long moment, head erect, eyes looking out over the crowd across the street. Then, as if she had been counting, she stepped down at a deliberate, processional pace to her car, still looking straight ahead, seemingly steeling herself. As others came out, Ted Kennedy loped up the stairs and inside again, and in a moment returned to the first family car. He stuck his head in the window, said, "I'm staying," and entered the cathedral again to join the long night's vigil over his brother.

After seeing the casket safely to St. Patrick's, the campaign team resumed its labors in Smith's office on the thirtieth floor of the Pan American Building at 200 Park Avenue, over Grand Central Station. The invitation list was coded to indicate categories of guests (K for family, S for Staff, F for friends, WH for White House, V for VIPs, D for Diplomats, M for others) and to indicate who would be invited to ride the train. Telegrams of invitation for the train bore the code T and later had to be shown to gain admittance. All through the next day the minute details were worked out for this most elaborate of public-private funerals—in some ways larger and more involved than the national mourning four and one-half years earlier that saw Robert Kennedy's brother laid to rest.

Outside the cathedral, police with bullhorns informed the crowds that St. Patrick's now was closed and would not open until 5:30 A.M. They urged the onlookers to go home, but many stayed. The line began forming—a line that was to grow, before dark fell again the next day, to the most incredible outpouring of human sorrow ever demonstrated on the streets of that great city. All along Fifth Avenue and nearby streets that night, shopkeepers already had stripped their windows and had placed black-bordered pictures of Robert F. Kennedy in them. When Robert Kennedy ran for the Senate from New York, they had called him a carpetbagger. Now in death, New York grieved for him as it would have for a son who had lived there all his life. And in the next two days it became clear, if it was not already so, that the nation—cities and towns where he had been, and there were thousands of them, and those where he had not—shared that grief.

By the time the sun rose on New York on Friday, June 7, a day just as punishingly hot and humid as the day before, the line outside the cathedral already had stretched several blocks and the wait was being estimated variously at from two to five hours. Even for a city that never sleeps, the turnout was astounding, and as the sun moved up and steamed Manhattan's giant concrete caverns, it grew and grew and grew. Police eventually were obliged to wind the line down two blocks to Park Avenue, then back to the Pan American Building at

West 45th Street and up the other side of Park toward the cathedral. Red Cross mobile units served soft drinks and water along the route, and fire hydrants were opened for drinking water as the temperature approached 90 degrees.

They all came, Robert Kennedy's special constituency, to say good-bye: the blacks, as they prefer to be called now; the old immigrant groups known in the shorthand of politicians as "the ethnics"; the poor and the nearly poor; finally and above all, the young. The expressions were uniformly solemn as the broad column of humanity inched forward like a huge snake, up one street and down another. He never shared poverty with those who suffered its hardships, but somehow their hardships got through to him, and they felt it. Nor was he as close in age to his young followers as he looked. Yet they too felt he was one of them, and they came to convey that—high school and college students, office workers and shopgirls, young mothers with small children. But the elderly were there too, some standing on crutches, and even many well-dressed business-men, who certainly were thought to be well beyond his appeal. They all stood quietly, respectfully, in this city of chronically complaining people, until after waits of two, three, four hours or more, they were ushered quickly out of the boiling, bright June sun and into the stone coolness of the nation's most famous place of worship.

Once inside, they found a scene of simplicity and splendor; the cathedral's soaring dark columns at the rear, then bathed in bright light at the transept, where six high golden candlesticks, and six members of the honor guard, flanked the simple African mahogany casket. Television cameras were perched on platforms tucked close to the columns. In the earliest hours, and periodi-cally throughout the day, the networks focused on the scene, capturing the crowds as they moved quickly by the coffin, as many as seventy-five to one hundred persons a minute. The heat from the television lights broiled the honor guard; it was changed every ten minutes or so. The nation's great and famous who took one or more turns that day were easily identified by those who walked down the aisle or knelt in prayer in the filled pews: former Sec-retary of Defense McNamara, UN Ambassador Goldberg, singer Andy Wil-liams, actor Sidney Poitier, civil rights leader Charles Evers, astronaut John Glenn and many, many others from every prominent walk of life. Others stood with the famous, or in groups of their own: friends and campaign workers

like Jim Whittaker, Rafer Johnson, the intrepid yet gentle Bill Barry, Ed Guthman; newsmen who covered the campaign: Warren Rogers and Stan Tretick
of *Look*, columnists Joseph Kraft, Art Buchwald, Rowland Evans, and Jimmy
Breslin, Gorey and Harwood and many others known only to their colleagues
and to politicians. The bittersweetness of the human warmth that flowed endlessly through St. Patrick's was accompanied by the whispered good-byes as the
mourners passed the casket: "God bless you," or "Rest in peace, Mr. Kennedy,"
or simply, "Good-bye, Bobby." As the honor guard members stood erect under
the blazing lights, all they could see were the hands—white and black, clutching rosaries and handkerchiefs—that reached out to touch the coffin for an
instant or to make the sign of the cross over it.

In midmorning, a woman faltered, fell as she reached the casket and cried
out, over and over, "No, oh, no, no, oh no." Police gently led her away. At one of
the front altars, another woman knelt, swooning, and moaned, "Oh, no, Bobby,
my Bobby." But for the most part, grief was borne well; many girls held hands
in line; some carried prayer books.

During the day, eight Masses were held, and many of Robert Kennedy's
fellow Catholics took Holy Communion in his memory, kneeling at the rail
past the coffin, just as he himself had taken Holy Communion that previous
Sunday, with Ethel at St. Mary's on the edge of Chinatown in San Francisco.
More than his brother, the late President, Robert Kennedy had been a "prominent" Catholic, and his presence at Mass on Holy Days in downtown Washington had been routine. Government secretaries often came to gawk at him in
church, and to fulfill their Mass obligation at the same time. During the campaign, he had gone to Mass every Sunday, sometimes at a little church nearest
to his hotel, occasionally going some distance to worship at a larger one with
an ethnic congregation of his special constituency. Most of the time, though,
he preferred to slip in and out without attracting attention or disturbing the
Mass. Now the Masses went on in the presence of his closed casket, and it was
all for him.

Members of the Kennedy family moved in and out of the cathedral all day.
Mrs. Rose Kennedy came once, knelt in a pew and shared a prayer book with
a soldier. A family group, led by Ethel Kennedy, composed as always, arrived
at about noon and knelt to pray in a pew across from the casket. The slight

murmur that had been floating through the church was stilled at the recognition of her presence. The line was halted temporarily, then allowed to proceed. Son David touched the coffin as the family left. The two oldest sons, Joseph and Robert Jr., stood honor guard several times, and on at least one occasion so did David. Jacqueline Kennedy came twice, once shortly after noon and once at about 4:30 P.M., this time with her daughter Caroline and son John. All three knelt at the foot of the bier, crossed themselves, touched the coffin and left by a side door.

Shortly after this visit to St. Patrick's, a related scene unfolded in another, contrastingly obscure place of worship clear across the country. Under heavy guard, the suspect in the assassination was brought from the Los Angeles Hall of Justice to the chapel of the Central County jail and there charged in one count with the murder of Robert Francis Kennedy. The room was selected because it afforded tight security. Standing behind a wooden altar, Superior Judge Arthur L. Alarcon called for "Case Number A-233421—The People versus Sirhan Bishara Sirhan." The accused, wearing blue jail dungarees and black patent leather slippers, was pushed into the chapel in a wheelchair, his left ankle supported by an ace bandage. Only when Judge Alarcon mispronounced his name did he speak. "It is Sear-han, not Sir-han," he said.

And at Arlington, McNamara had spent the afternoon at the Kennedy gravesite, inspecting possible locations. Three spots were picked, to give Mrs. Kennedy a choice, but one, about 20 feet from President Kennedy's grave, was preferred by McNamara. After discussions with the widow later that night, it was selected. As Defense Secretary under Kennedy, McNamara had made the burial plot a matter of his most personal concern. He envisioned it as a lasting memorial not only to the late President but to the high purpose of government service he inspired, where key members of the Kennedy Cabinet—including himself—could be buried someday.

Back in New York, St. Patrick's was to have been closed at 10 o'clock this night, but when it became apparent those joining the line in the late afternoon would never gain entry, the family decided to permit the viewing to go on through the night. By 5 o'clock the next morning, thousands more had walked through, and still the line stretched out. At the Kennedy apartment at UN Plaza, a gathering akin to an Irish wake—good cheer pushing out the grief—filled these hours. Then the Senator's wife, accompanied by Barry, came to St. Patrick's for a final moment alone with the casket, for from that time until Arlington, all would be in public view. New York police and church officials quietly asked the hundreds still inside to leave, and as Ethel knelt and prayed, they departed. The police too left, leaving Ethel alone with her husband, with only the watchful Barry standing by.

Through the night, Mankiewicz, Salinger and other staff members had labored to complete the logistical arrangements for the next day. Telegrams went out to the 2,300 guests, and well before 9 o'clock Friday morning the first of the invited began to arrive and to fill the giant cathedral. Each received a Mass card bearing Kennedy's likeness—a particularly happy and boyish picture was chosen—and quotations selected by Walinsky and Edelman. More thousands lined Fifth Avenue across the way and on down for blocks and blocks toward Pennsylvania Station on West 34th Street, where the funeral train awaited. Thirteen active pallbearers had been chosen: brother Ted; son Joseph; brother-in-law Steve; McNamara; former Secretary of the Treasury C. Douglas Dillon; Ambassador W. Averell Harriman; Lord Harlech, former British Ambassador to the United States; General Maxwell Taylor; John Glenn; Jim Whittaker; LeMoyne Billings; Dave Hackett and John Seigenthaler. Forty-nine were honorary pallbearers, including Ted Sorensen, Larry O'Brien, Dick Goodwin, Arthur Schlesinger, Bill vanden Heuvel, Fred Dutton, Stash Radziwill, Charles Evers, Cesar Chavez, and brother-in-law Sargent Shriver; and twenty-three were to be ushers: Barry; William Brady; John Burns; Joe Dolan; Dutton; Peter Edelman; John English; Paul Fay, Undersecretary of the Navy under John F. Kennedy; Joseph Gargan, a cousin; Rosey Grier; Guthman; Rafer Johnson; Tom Johnston; Mankiewicz; John Nolan; O'Donnell; Barrett Prettyman; Rushton and James Skakel, Ethel's brothers; Salinger; Ben Smith, a family friend and former Massachusetts Senator; Frank Thomas of the Bedford-Stuyvesant community project in Brooklyn; Walinsky.

Among the early arrivals were Harriman, McNamara, Dillon, evangelist Billy Graham, civil rights leader Ralph Abernathy and Glenn, all of whom stood honor guard. So did Vice President Humphrey when he arrived, about twenty minutes before the 10 o'clock Solemn Requiem Mass started. The front pews were filled with other notables: Senator Eugene McCarthy, talking with the 1964 Republican Presidential candidate, Barry Goldwater; Chief Justice of the United States Earl Warren; Secretary of State Rusk; Governors George Romney of Michigan and Nelson Rockefeller of New York; Mayor Lindsay; Chairman of the Joint Chiefs of Staff General Earle G. Wheeler; former Vice President Richard M. Nixon; House Speaker John W. McCormack and a host of Senators and Congressmen. About ten minutes before the service was to start, President Johnson strode in, somber-faced, past the casket to a seat in a front pew, to the left, on the side opposite the Kennedy family. The guests rose, then sat when he did. The President, in a dark gray suit, watched as the Kennedy sisters took their places. When Mrs. Jacqueline Kennedy entered from the side, the guests rose again, and then a third time when Ethel and Ted Kennedy took their places in the front row, right, with the older children of the late Senator.

For the next hour and forty-five minutes, the reds and maroons and purples of the Catholic hierarchy filled the cathedral, offering the kind of solemn yet magnificent departure reserved in past centuries for kings and others of the mighty. Yet nothing so captured the mood and the grief of the assembled mourners than a short and unannounced eulogy by Edward Kennedy. It was a labor of love written by Walinsky, with help from Gwirtzman and Burke, and drawing heavily from the words of the deceased. Looking down on the flag-draped casket in front of him, the one surviving brother of four said of Robert Kennedy: "He gave us strength in time of trouble, wisdom in time of uncertainty, and sharing in time of happiness. . . . He loved life completely and he lived it intensely." Amid all the splendor of his own ancient church and its ornate, soaring cathedral, Edward Kennedy said simply of him who had died: "My brother need not be idealized, or enlarged in death beyond what he was in life, to be remembered simply as a good and decent man, who saw wrong and tried to right it, saw suffering and tried to heal it, saw war and tried to stop it."

The young Senator's voice broke as he said this, but he persevered, ending with the same words with which his brother had concluded so many of his own speeches across the nation: "Some men see things as they are and say, 'Why?' I dream things that never were and say, 'Why not?' " For the men and women who heard the eulogy and who saw Edward Kennedy struggling with his own emotions, there were different peaks of poignancy. But for those who had followed the campaign trail with Robert Kennedy over the previous eighty-four days, and particularly for the newsmen to whom the line had special significance, none so brought home to them the personal quality of their loss than this last reference to Shaw.

Archbishop Cooke, in white robes, took the pulpit and counseled that "the act of one man must not demoralize and incapacitate two hundred million others. To permit this to happen would be to fail utterly to grasp the message of hope and optimism in Senator Kennedy's life." It was a strong sermon, but this was a day in which what was seen counted for more than what was said. Senator McCarthy, he who had grown so vocally bitter about Robert Kennedy during the campaign, dropped his head in his hands several times. Leonard Bernstein, leading thirty musicians of the New York Philharmonic Orchestra in a somber movement from Gustav Mahler's Fifth Symphony, weaved back and forth as he conducted, as though he might topple; and two other widows there who had been victimized by assassins' bullets, Jacqueline Kennedy and Coretta King, sat mournfully and endured this repetition of a personal agony all too familiar.

Perhaps what was most moving, though, more so even than Ted Kennedy's eulogy, was a procession of eight small Kennedy children—sons, daughters, nephews and nieces of the deceased. They lined up, two by two, at the head of the casket and carried to the altar the Host and the wine used in preparing for Holy Communion. The girls in white and in blue, the boys in jackets, except for John F. Kennedy, Jr., they marched without a falter up and back again, then took their seats with their parents.

Near the close, over the stillness of the huge Gothic church, the familiar voice of Andy Williams—without accompaniment, firm yet gentle, and hauntingly slow—blanketed the assemblage, giving to "The Battle Hymn of the Republic" a measured beauty and sadness that for many it had not had before. At the end of the third chorus, the congregation, the cathedral organ and chorus joined in a mighty crescendo. Richard Cardinal Cushing, the seventy-two-year-old Boston prelate who presided over John F. Kennedy's burial, anointed the coffin and said, "May the Angels take you into paradise; may the martyrs come to welcome you on your way, and lead you into the Holy City. . . ." Then it was over; President Johnson rose, walked slowly past the front row and spoke briefly to

Ethel and other members of the family, including Jacqueline Kennedy. Then he left by a door behind the altar, Secret Service agents packed tightly around him.

To a rousing choral rendition of Handel's "Hallelujah Chorus," Ted Kennedy and young Joe led the procession down the long aisle and onto Fifth Avenue, with Ethel, the children and the rest of the Kennedys behind the casket. Ted helped carry the casket to the waiting hearse, took the flag off it as it was put inside, then joined Ethel to ride with the body. Again, as she had done two nights before, Jacqueline Kennedy appeared on the top step of St. Patrick's, still beautiful in her own regal style of grief. She stood there for a very long moment, wan and brittle, drawing all eyes, then walked to her car with her two children and her sister, Lee Bouvier Radziwill.

The short trip to Pennsylvania Station was a marvel of logistics. A fleet of thirty buses was parked on 51st Street, and about 700 invited guests were directed quickly to them. The silent crowd stretched for more than twenty city blocks, three and four deep at some intersections, as the cortege moved down Fifth Avenue. Guests were funneled to an escalator at Penn Station that provided the Secret Service and other security men a vantage point for checking credentials. Those who did not have exactly what was required were turned away or subjected to a search before being permitted to go on. Earlier, at St. Patrick's, a man had been arrested for having an unloaded gun in his briefcase. The police were determined this tragedy would not be compounded because of their carelessness. It was not; at least not for that reason. But more heartbreak was ahead nonetheless, just down the tracks in New Jersey.

The twenty-one-car train pulled out of Penn Station at 1:04 P.M., thirty-four minutes behind schedule—not bad at all for a Kennedy trip slowed as usual by huge crowds. The guests and press occupied all but the last five cars, which were reserved for the family and the closest friends and associates. The casket was placed in a special observation car, whose rear platform was trimmed with garlands of fresh green leaves. One of two picture windows, one on each side of the car, was removed for loading the casket, and then replaced. When it became evident that the casket could not be seen very well from the outside, it was placed on chairs. Throughout the long journey, friends again stood honor guard, making sure at the same time the coffin did not slip off its makeshift perch. Ethel Kennedy remained in the last car much of the time, or in adjacent

cars. There was some quiet weeping among the family, but mostly the same remarkable stoicism that the nation had come to expect from this clan in which
the virtue and the strength of self-discipline was an article of faith.

As the 226-mile journey began, the mood throughout the train was one of
sober anticipation. Robert Kennedy was being taken to meet the people for the
last time, and there were few aboard who did not expect an outpouring of mass
emotion along the tracks to match anything they ever had experienced. Almost
at once, they knew they were not wrong. As the train crossed a bridge over the
Passaic River, a Newark Fire Department tugboat, with its name printed in clear
red letters—JOHN F. KENNEDY—floated in silent tribute, its crew on deck. And
as the train pulled into the station at Newark, an immense crowd of about 7,000
jammed the narrow pavement along the track. It was perfectly silent, except for
the click-clicking of the wheels as they rumbled slowly by. Expressions were
somber, and many wept. There were a great many small flags, waved ever so
slightly in a gesture of good-bye, and everywhere there were black faces—old
and wrinkled and crying, young and tearful. Men pulled themselves tall—you
could see it happen as the train moved by—and some placed right hand over
the heart, or saluted. There were simple signs, like one that hung around a
young girl's neck and said, misspelled, FAREWEL, ROBERT.

At 1:24, the train passed the Elizabeth station, where the crowd and its reaction were much the same. Some, in their grief, ignored railroad warnings
and jumped down onto the tracks just to touch the side of the observation car.
Many along the route had touched him as he campaigned over the last years
and months, and now this was the best they could do to show how they felt.
This time, because they did, tragedy was compounded. A four-car train, the
Admiral, bound for New York from Chicago by way of Philadelphia, roared
past; passengers on the Kennedy train looking out at the trackside mourners
were horrified to see some of them turn and scamper for safety. Two of them—
John Curia and Antoinette Severini, both fifty-six years old—did not make it
and were crushed under the wheels of the Admiral. Six others, including a
three-and-one-half-year-old granddaughter thrown to safety by Mrs. Severini,
were injured.

Word of the tragedy swept swiftly through the Kennedy train, at first only
in whispered snatches, finally like a heavy shroud over all discussion. Ethel

Kennedy apparently did not know, and was not told. Jerry Bruno, as the chief Kennedy advance man, contacted railroad officials and told them he wanted an absolute guarantee that no more trains would pass the Kennedy Special until it reached Washington. When, according to Bruno, they balked, he said: "Then we'll stop this train right now." Finally, the officials agreed. The railroad insisted it had invoked tight security, but every station on the route again was contacted to apply strict measures along the tracks, and the Kennedy train itself slowed. At one point, it stopped for inspection of a smoking brake shoe, then moved on. But the tragedy was not at an end; at Trenton, eighteen-year-old Joseph Fausti climbed on top of a boxcar for a better view, brushed against a high-voltage overhead power wire and was critically burned. This news too swept the train but was kept from Mrs. Kennedy.

According to the day's schedule, Senator Kennedy's body was to arrive in Washington at 4:30. But by that time, as a result of the mishaps and the tremendous crowds, it was only approaching the North Philadelphia station. At each depot, crowds jammed the track platform; they also stood on overpasses, on green banks along the route, at open windows in tenements and on rooftops. Local high school bands turned out to play "The Battle Hymn of the Republic" or "America"; men and women held their infants and small children high so they could see; fathers and sons in shorts or swimming trunks, stripped to the waist in the hot sun, stood holding fishing rods or baseballs and gloves. Many placed their ball gloves over their hearts, or put their hands to their baseball caps in salute; some knelt in the grimy roadbed, the coarse gravel cutting into their bare knees. And all along the route, the preponderance of black faces, and of faces with their immigrant stock plainly written on them, was undeniable; so was the particular grief their expressions confessed. They told with wordless eloquence what Robert Kennedy had meant to them, in their private battles against discrimination and poverty, and how they measured their loss.

As this lonesome train with nearly a thousand people on it moved with excruciating slowness down the track in the punishing heat, many who had campaigned with Kennedy thought now of all the other train rides—the buoyant spirit aboard, the excited faces that lined the routes. They were, quite obviously, the same kinds of people on this day who had turned out in those earlier, happier times. It brought back in painful clarity one train ride especially—the

Wabash Cannonball through Indiana in those difficult but challenging cool April days when the campaign of the Senator from New York was undergoing its first real test. Ethel and Bob Kennedy stood in the parlor car as they listened to the reporters' parody, and they grinned with all the rest at the line contributed by Dave Halberstam: "The blacks in Gary love him, the Poles all fill his hall; there are no ethnic problems on The Ruthless Cannonball." Now as then, there was more truth than poetry in that line.

As the train rolled toward Washington, young Joe Kennedy, the eldest son at fifteen, worked his way through the entire twenty-one cars, discarding his usual reticence and introducing himself to every single passenger: "Hello, I'm Joe Kennedy. Thank you for coming." And to those who offered condolence, the awkwardness of his age would reveal itself in his answer: "That's okay."

A little later, when the train was in Pennsylvania, the word spread, back to front, that Ethel Kennedy was coming through. To some it seemed incredible, and then, on reflection, only natural. It was her way; in her direct, uncomplicated manner, she always had been the one to break the tension. Bill Barry moved protectively ahead of her, and as she approached, men sprawling in their seats put their drinks down and their jackets on. The young widow shook hands with every passenger, thanking each and keeping it light. To many she had a special message, putting her arm around them as if they were the most bereaved, embracing and kissing some on the cheek, greeting them as if they had just called on her at Hickory Hill. Four and a half years earlier, Jackqueline Kennedy in her grief had displayed a stolidness and queenly determination that infused all those around her with strength and admiration; now, this outgoing, cheerful young woman achieved the same result, in her own natural way.

After Ethel had passed through the press car, Frank Mankiewicz came in, stood on a seat at the front, and presented a full report on the accidents at Elizabeth and at Trenton. He roasted the Penn Central Railroad officials, saying the Kennedy staff had "made it very clear" that security of onlookers along the route needed special attention. Throughout the Kennedy campaign, the matter of safety—for the spectators, not just for the candidate—had been a constant cause of worry among the Senator and his aides. They feared in every motorcade that some small child or elderly woman would be swept under the wheels of a car or dragged along, so closely did the crowds press against the cars and run along with them. Yet in the previous eighty-four days, no such serious accident had occurred. And then, on the eighty-fifth, this.

The briefing was interrupted when Ethel Kennedy started back toward the family cars. She still did not know about the accidents, and when she reentered the press car and found Mankiewicz holding forth, she grinned. "You're kidding," she said. "It's too hot in here for that kind of thing. What are you talking about?" The reporters laughed, and passed it off, until she had moved out into the next car. Then the grim report proceeded.

At the North Philadelphia station, another huge crowd sang "The Battle Hymn of the Republic" as the train eased by, nearly three hours late. The contrast between the deep, unfettered grief outside and the growing casualness inside had a sobering effect on many of the passengers. It was emotionally difficult to live with intense sorrow hour after hour like this, yet the reminder at every stop of what this train and its fallen passenger meant to those outside demanded it. Ted Kennedy was so moved that he went to the rear platform and waved his thanks to the crowd.

From Philadelphia to Wilmington to Baltimore, a trip that usually takes about ninety minutes and on this occasion required nearly three hours, the long day finally began to take its toll. The somberness that had slipped to casualness en route became a kind of weary insensitivity; passengers jammed the club cars, demanding drinks and griping for lack of ice and food. Conversation became louder and lighter, girls in black mini-dresses tripped up and down the aisle, and a frivolity took over in many parts of the train though not, from most reports, in the family cars. By the time the train moved into Washington's Union Station at 9:25 P.M., eight hours and twenty-one minutes after the New York departure, the beginnings of an old-fashioned Irish wake were evident.

But for those who knew Robert Kennedy best, and they really constituted only a small percentage of those who were fond of saying they knew him, the marathon train ride was an ordeal that tested their endurance. While others played it light, O'Donnell and O'Brien, Barry and Mankiewicz hung in and waited for it to end. John Seigenthaler, commenting near the end about the interminable pace, suggested that if the trip had lasted much longer, Robert Kennedy—not one to suffer delay willingly—would have started "kicking the box." It was a remark that only a close friend could make, and only other close friends could appreciate.

At Union Station, the crowds lined the route from trackside out into the street and on up to Capitol Hill. The family cars had been uncoupled and brought in on another track, to facilitate loading of the casket into the hearse that would take it through the city to Arlington. Most of the passengers from the train were herded into buses that took them directly to the burial site. A small motorcade was formed, including only the closest relatives, friends and the highest government officials, including President Johnson and Vice President Humphrey. One press bus, for only those newsmen who regularly had covered the Senator's Presidential campaign around the country, was insisted upon by Mankiewicz, not as a pool, but as a symbol of Robert Kennedy's close ties with these traveling companions.

Here after eighty-five days, was the final motorcade in the campaign of Robert Kennedy. In some ways, it was like all the others—the huge throngs lining the streets for a glimpse of him who was to have brought back the past again, or was to usher in a new future; always that mix of memory and promise. In the nation's capital as elsewhere, black faces on this night were plentiful, but there were many whites as well; Washington had been Robert Kennedy's home for more than sixteen years, and it had been one of his special interests that offered little potential gain, like the Indians; as Cabinet member and Senator, he had helped bring to the city an excitement and accent on youth that it had not known for years, and now was not likely to know for many more.

The motorcade moved at an even pace up from the station past Kennedy's Senate office and the Caucus Room where he had begun his Presidential campaign eighty-five days before. The crowds were silent, thicker in some places than in others, after a long vigil marked by intermittent drizzles and threats of worse weather. Some along the way held candles; more held matches, which they struck hurriedly as the motorcade approached. The hearse and the cars behind it swung down Constitution Avenue, past the Justice Department where Robert Kennedy was Attorney General for more than three years. The motorcade rolled past the ellipse and the White House beyond, the destination he had set for himself when it all began; it moved on toward the Lincoln Memorial and, across from that, Resurrection City of the Poor People's Campaign then in progress.

As the funeral party moved along Constitution Avenue, the press bus, under direction of the usual Kennedy lieutenants, was adhering to the campaign dictum this one last time: stay up with the candidate. The bus had been assigned twenty-third place in the line, but Mankiewicz ordered the driver to swing out and up to about six cars from the hearse, just behind the family. It soon became clear that certain VIPs—apparently both the President and Vice President—were thus made to suffer the indignity of trailing the press bus, and a Secret Service agent was dispatched to pull it out of the line. As the agent barked orders to the driver, Dick Tuck from within the bus suddenly blew a whistle, ordered the driver to proceed, and left the agent standing aghast in the street. Limousines were passing the bus now, so Dick Drayne jumped out and planted himself in the middle of the motorcade. As he had done in countless others, he stopped the next approaching limousine and began to wave the bus back into the line. It turned out, though, that among those seeking to pass was one carrying Lyndon B. Johnson, President of the United States. A Secret Service man opened a car door and waved menacingly until Drayne, grinning broadly, finally broke off and swung aboard the bus again to cheers and applause from its passengers. Those who lined the streets in such silent reverence no doubt would have been shocked had they known of the raucous behavior on the press bus. But in its way too, it was a sentimental gesture—a spontaneous effort to hold on to what had been, because it was something of value.

The motorcade halted at the Lincoln Memorial for still another choral rendition of "The Battle Hymn of the Republic" and in special recognition of Resurrection City and its inhabitants, many of whom felt this was the only white politician who had listened and had understood. Then the cars moved silently and alone across the Memorial Bridge, which had been cleared of all traffic and all crowds. Up ahead, just below the Lee-Custis Mansion, where the eternal flame marking the grave of John F. Kennedy usually could be seen, the lights of television crews marked the place where the long journey of Robert Kennedy was to end.

The cars and the bus climbed a winding hill to a roadway just forty yards or so below the Kennedy gravesite. Spectators lined the way, holding lighted candles or matches. The pallbearers who had carried the casket from St. Patrick's Cathedral now brought it up the walk to the gentle sloping hillside where the forty-two-year-old Senator would be laid to rest, only twenty feet from the grave of his brother. Ethel Kennedy, escorted by brother-in-law Ted, watched and took her place to the downhill side of the slope. Other members of the immediate family moved into place, and the President and Mrs. Johnson walked off beyond the head of the casket, a fair distance away. All other notables, including Vice President Humphrey, were bunched around down the slope, behind the Kennedy family or off to the other side. Many of these, too, held candles. The crowd later was estimated at more than 1,000.

Now, four years, six months and thirteen days after their first eventful trip to Arlington, the Kennedys were all on the same peaceful spot overlooking the Potomac and Washington, here to bury another of their own, victimized by an assassin. This time it was to place the third son near the body of the second under circumstances so bizarrely similar that now, even after the four days of living with the tragedy, it all seemed a bad dream.

There were, of course, some striking differences. This time it was at night, about ninety minutes from midnight, and the combination of a nearly full moon and the glare of floodlights somehow added to the unreality of it. The setting itself was more beautiful. On November 25, 1963, John Kennedy was buried on a plot with artificial grass around the base of the coffin. In the intervening time, a permanent memorial site had been constructed, and fresh young trees now in full bloom of the hot spring evening removed some of the starkness of that earlier burial. This time, the music was simple—"America the Beautiful" by the Harvard University Band—and the honor guards were civilians. But one quality remained: the Kennedy tradition for grace under pressure. None of the large family gave way to tears—not the young wife left with ten children and expecting an eleventh; not Mrs. Rose Kennedy, the mother who had two sons assassinated and two other children killed in accidents; not the sisters, Pat, Jean and Eunice. And once again, as on that memorable November afternoon in 1963, Jacqueline Kennedy mastered her grief and watched without tears. Even the small children, who followed their mother and their aunts and uncles to the African mahogany casket and kissed it, maintained their composure. Once, young Joe was seen to brush the corner of his eye. But a moment later, when one of his young sisters came to his side after kneeling at the casket, he reached behind her and rubbed her shoulders and back to soothe her.

Patrick Cardinal O'Boyle of Washington said the parting prayers and then the flag was quickly folded in the traditional triangle by the pallbearers. John Glenn, the first American to orbit the earth, took the flag, walked several steps over and handed it to Ted Kennedy, who turned and gave it to Ethel at his side. There was total quiet and attentiveness in the crowd; hot wax dripped from candle to hand, stinging for a long moment and then crusting on the fingers. But no man or woman or child cried out or stirred.

In the presence of the stricken family, the other notables, including the President and Vice President, went almost unnoticed. The Johnsons, at the close of the ceremony, stepped forward, said a few whispered words to the widow and other family members, and disappeared into the warm night. In the somber mood, there was no trace of the bitterness that often had existed between the Kennedys and Lyndon Johnson. The President was given a place of honor at graveside, as he had been at St. Patrick's at the start of this long last day.

Off to the opposite side, under a tree and hardly noticed, was Princess Grace of Monaco, the former American movie star Grace Kelly, her eyes brimming with tears. Secretary of State Rusk, whose position on the Vietnam War Robert Kennedy had so vigorously opposed but whom he declined to attack personally, stood far down the hill behind some reporters. Secretary of Agriculture Orville Freeman peered over the shoulder of Vice President Humphrey; other Cabinet members were lost entirely in the hushed crowd. Someone held Freckles, the Senator's springer spaniel, as he strained at his leash, wanting to join the family.

Ethel Kennedy, clutching the flag, and Ted Kennedy walked to the casket, knelt and kissed it, then moved off. Other family members and the close friends then climbed up the green slope to say their own good-byes; it had been a punishingly drawn-out day, fourteen to fifteen hours since most of them had entered St. Patrick's, but still it was hard to break away. It had been, after all, less than four days since Robert Kennedy's bright dreams for political leadership and domestic reconciliation in a time of national crisis had been shattered in the kitchen of the Ambassador Hotel, 3,000 miles away. And now he was at rest, close by the grave of his brother, in whose footsteps he always had wanted to follow, and never had made any secret of it. Jacqueline Kennedy, after saying her farewell to Robert, led her daughter and son the 20 feet to her own husband's grave, and gave each of them a sprig of flowers from Robert's burial to place on their father's grave. It was an unscheduled gesture—and one that did not sit well with some other family members.

For long minutes, many in the crowd stood holding their lighted candles, looking at this newest gravesite, with the floodlights casting long soft shadows over it. They moved up the slope to the casket, some kneeling, some reaching out to touch it, and then slowly, reluctantly, they would break away and walk down the hill to the waiting cars and buses, still holding the lighted candles. The feeling of not wanting to let go seemed to pervade the scene; many walking to their cars turned back for a last look, and then again. Some milled around, as if not knowing even after this terribly, painfully long day, what to do next. The regular reporters of the Kennedy campaign climbed aboard their bus, the jovial mood now thoroughly dispatched by the final burial scene. They sat silently as the driver took them back into Washington. There were on the bus wide differences of opinion about Robert Kennedy, candidate and man. But they seemed to fall away in this last motorcade, this time without a candidate and with no need to stay up close to the lead car. The bus crossed Memorial Bridge into downtown and discharged some passengers near their offices. The rest were going on to the Shoreham Hotel, where their baggage was waiting. As I rose to get off, Chuck Quinn of NBC, who had been with Robert Kennedy almost solidly for twelve weeks, looked up with a pained expression and said quietly, "Aren't you going to the end with us?"

The same desire to hold on led many of the close family friends to follow Ethel Kennedy home from Arlington to Hickory Hill, just ten minutes away up the Potomac. There, where they had shared so many lively times, as many as a hundred mourners stood around in the dining and living rooms on the first floor, having a drink and a snack. The McNamaras were there, and the Dillons and Harrimans; the Glenns, the Hacketts, Rafer Johnson, Andy and Claudine Williams and many, many more. It was, as always at Hickory Hill, a mixed bag, the kind Robert Kennedy enjoyed. The widow joined in, moving about and talking, until the last of them left, as late as 4 o'clock Sunday morning. And even then, down the river at Arlington, about a hundred mourners waited below the long gentle hillside where the Kennedy brothers now rested, for the hour when the cemetery's high iron gates would be swung open again. It was over for Robert Kennedy, but as with John Kennedy, it would go on.

CHAPTER TEN

The Letting Go

CAMELOT GOES ON

IN THE DAYS and weeks that followed, the experience of the three-month campaign of Robert Kennedy became, in retrospect, singularly precious for many and they hoarded its memories and were bound together by their common loss. In the first days, at Hickory Hill, many of those who had been drawn to the man streamed in and out in genuine solicitude for Ethel. To no one's surprise, the incredible ability to bounce back that living with tragedy had brought the Kennedys demonstrated itself. On the day after the burial, as an estimated 60,000 mourners filed past Robert Kennedy's grave, some of the sisters played tennis at Hickory Hill and went swimming in the pool with Rosey Grier, Rafer Johnson and other guests. It was all part of the business of going on. In a few days, Ethel and the family went to Hyannisport for the summer; there they swam and sailed and mended their lives again, as they always had, in a surface of flurry and activity and a private accommodation with life that had so blessed and punished them. Friends were always there, especially those who had been involved in the last campaign, and there was good talk and the athletic life. The older children went off for the summer to camp, and to Europe, and to Africa. It was, for the Kennedys, the only way.

At the Kennedy campaign headquarters at 20th and L Streets Northwest in Washington, where many had not tasted a Kennedy tragedy before, the dismantling was achieved amid a mood of clinging to what was and what might have been. Boxes were packed and posters and pictures taken down, but the sense of organization and belonging remained and was nurtured. Young Joe

Kennedy, dressed in a white shirt, old bright blue trousers and scuffed loafers, poked around the boxes and the stacks of papers, as if searching for aspects of his father that he had not known, had not had time to find out about.

Up on Capitol Hill, in the Senator's office, it was the same. Because Senate rules gave the staff two months to attend to unfinished business, Dolan, Mankiewicz, McDonald, Walinsky, Edelman, Angie Novello and others continued at their desks daily, pursuing favorite Kennedy projects, working on various memorial ideas, helping other staff members find new jobs, make new plans. As they did all these things, the boxes were packed and the cherished photographs were taken from the walls, so that the degree of emptiness in the offices measured the time still to go before this part of the Robert Kennedy operation would have to end and they would have to let go. One by one, as the deadline approached, they left; many, like Mankiewicz and Dolan, just got up from their desks one day and went out, without good-byes. Only six younger staff members were there at the end when the door was closed and the sign MR. KENNEDY, NEW YORK, WALK IN was taken down.

But these last two months were just a prolonging of what was over. For some, the letting go came four days after the burial, at a staff party held in the home of Carol Welch, one of the Senator's traveling secretaries. There was little gloom, much talk about the Senator and the campaign, and the same sense of elite togetherness that always existed in the Kennedy camp. The stereo set thumped Burt Bacharach and the voices chattered over it. There was guitar playing and singing, by Walinsky and others, as of old. The wanting to go on was obvious, yet there was no talk then of trying to put it together again, under somebody else, not even Edward Kennedy. These were issue-oriented people, but most of them were, more than that, Robert Kennedy oriented people; for them there was no sense in talking about more politics in 1968. Dick Tuck, known mostly for his irreverent humor, said to a friend: "Camelot is ended. It began in violence and ended in violence. It began with John Kennedy's death, not before, and it ended with Bob Kennedy's assassination. It will never exist again, whether Teddy ever gets to be President or not." It was, of course, a premature epitaph, but it reflected the mood at the time.

In the weeks that passed, in increasing speculation and pressures for Teddy Kennedy to accept the Vice Presidential nomination on the Democratic ticket, there seemed a failure among many to appreciate that Robert Kennedy had been more than a man who had tried to put his family back in power. Had the reestablishment of the Kennedy dynasty been all there was to it, perhaps Teddy would have grasped at the opportunity waved at him by several Democratic governors and Mayor Daley. But with Robert gone, the purpose and the compulsion for Teddy and for Robert's closest followers were gone too, for a time at least. Running with Humphrey, the man who represented the status quo on Vietnam policy that Robert had fought so vigorously, certainly could not be the answer.

There were those—like Sorensen, Mankiewicz and Salinger—who tried to salvage something politically by persuading Senator George McGovern to serve as a holding candidate for the Robert Kennedy forces at the national convention. But it was little more than that—that and a final signal to McCarthy that his treatment of Kennedy in the primaries had not been forgotten. The song said, "They're noted for long memories on The Ruthless Cannonball," and even without its "demon driver," that was true.

McGovern himself entered the race and conducted it in admirable good spirits and taste, without rancor toward McCarthy or anyone else. One word— "compassion"—ran through his short campaign, and it was the word that said more than any other why he and the Kennedyites felt the need for an antiwar alternative to McCarthy. As "law and order" and "crime in the streets" had become code phrases for anti-Negro sentiment, "compassion" was a polite way to say what the Kennedyites felt about McCarthy—that he was a cold fish, that he cared about humanity but couldn't be bothered with humans, especially the blacks and the other disadvantaged. It was a harsh judgment, and probably an unfair one, but McCarthy's public conduct, in which he made a studied effort not to show his feelings, encouraged it.

Also, McCarthy's jibes at Kennedy for courting the black and ethnic votes on the primary circuit, and particularly his slur in Oregon on the mentality of Kennedy voters, had cut deeply. But beyond all that, there was the feeling—and not confined to Kennedy backers either—that McCarthy did in fact lack a powerful emotional as well as philosophical commitment to the nation's poor; such a commitment, the McCarthy critics believed, was required to ignite public opinion, and through it any really meaningful government action.

For most of the old Robert Kennedy team, however, there was no stomach at this juncture for any more politics in 1968. The pro, Larry O'Brien, hopeful he could help stitch the splintered party together, went over to his friend Humphrey, and Goodwin returned to McCarthy. But the other key figures who didn't ride the McGovern vehicle sat out the preconvention doldrums. Perhaps in time their attitude would change; it had been the same way, after all, when John Kennedy died. Except this time there was the awful awareness that another quest for the top by a Kennedy, in a nation of guns, violence and instability, could invite—some even said could ensure—a third tragedy.

In any event, the shared association with Robert Kennedy in his last campaign bound many of the participants into a cult quite separate from the old John Kennedy political clique. One Saturday night in mid-August, about seventy of the Kennedy campaigners gathered in a private room at Duke Zeibert's Restaurant in Washington for a showing of hundreds of color slides of the campaign; intimate, inside-the-entourage snapshots that captured all its camaraderie. Mankiewicz, Drayne, Bruno, Barry, Dolan and a host of reporters and wives were there. For about two hours the slides flicked onto the screen, sparking high good humor and hilarious banter as the comrades in a great adventure relived it.

But near the end, on three separate occasions, slides of the Robert Kennedy funeral inadvertently were included and were flicked off immediately. The reaction in the darkened room each time was to act as though no one had seen the slide, and even, at the next campaign cameo, to laugh a little louder. To close the private showing, about sixty color slides of the candidate on the stump and in reflective moods were shown. They were brilliant studies by *Look*'s Stan Tretick of Robert Kennedy at his most impassioned, his most convincing and contemplative, and these last especially captured that sympathy-evoking quality that drew people to the man. A total stillness fell over the room, broken only by the periodic click-click of the projector as Tretick changed slides. For about ten minutes the face of Robert Kennedy filled the screen and the minds of all those there who had known him at close range; when the last slide was shown, Mankiewicz broke the silence by saying, simply, "Thank you," and the lights went on. Everybody got up, wordlessly, and began to move out. Again they were confronted with letting go, and it was not easy.

Part of letting go, among most of the Robert Kennedy followers, was the hope and the anticipation that they could ignore the approaching Democratic National Convention. It would be too painful for them, and it seemed almost irrelevant now. But when it came, with the Vietnam policy that Robert Kennedy had sought to end the dominant issue, ignoring what happened in Chicago was impossible. In all the pulling and hauling over the platform plank on Vietnam, the assassinated candidate's influence was ever-present; in the deep pessimism that infected the convention from the start and inevitably produced a yearning for a viable alternative to Humphrey and McCarthy, the prospect that he could have been nominated, had he lived, seemed infinitely more possible than it had appeared to be on that last night in California.

But Robert Kennedy had not lived, and those who were desperate to produce a candidate more likely than Humphrey to oppose Johnson's Vietnam policy and to save their own political skins in November did not have to look far for a replacement. Ted Kennedy, in his first public speech since his brother's death, had said only a few days before the convention opened that "like my three brothers before me I pick up a fallen standard." Although he also had said, for the second time since Robert's death, that he would not be available for national office in 1968, some yet hoped they could tap through him the support and the sentiment that Robert Kennedy in life and death had summoned.

Jesse Unruh and—to a degree not generally known in Chicago at the time— Dick Daley put pressures on Ted Kennedy in Hyannisport to indicate his availability for the Presidential nomination. A full-blown "Draft-Ted" effort bloomed simultaneously, with former Governor Michael V. DiSalle of Ohio the nominal leader. After several hours of suspense, Kennedy phoned DiSalle and asked him to desist. But behind the scenes, the pressures continued from Daley and Unruh. And as tempers mounted over the Vietnam plank and the prospect that Humphrey's nomination would mean a Democratic campaign based on more of the same in Vietnam, more of the old Kennedy hands began to stir. Maybe, in the favorite Kennedy phrase for the politically feasible, it could be "put together" to stop Humphrey after all.

Steve Smith had been dispatched to Chicago by Teddy to keep an eye on things and to protect the Kennedy interests. From the Standard Club, a private downtown sanctuary, he conferred with Unruh and other key antiwar

Democrats, and with Daley. Daley was pushing, but Teddy stayed cool, wary of the uncertain prospects and of being trapped by Daley into dropping his position of unavailability and being saddled with the Vice Presidential nomination. Also, in a meeting with Humphrey several days before, Teddy had assured Humphrey he would not become a candidate. As the antiwar forces buzzed, Smith received a phone call from Dick Goodwin, the old McCarthy-Kennedy-McCarthy aide, saying Gene McCarthy would like to see him. Smith along with Fred Dutton went to the twenty-third floor of the Conrad Hilton Hotel and met for about fifteen minutes with McCarthy and Goodwin. After talking around the situation in one of those typically obscure and stony sessions that by now had become part of the McCarthy-Kennedy feud folklore, McCarthy asked Smith to step into the bedroom alone with him. Words were exchanged, they came out again and Smith and the aide left. Shortly afterward, Smith told another Kennedy insider: "We can have the nomination." McCarthy, acknowledging he was going nowhere, had asked whether Teddy would enter if he stepped aside. If so, McCarthy had said, he would withdraw. But in order not to completely disappoint his followers, he wanted to have his name put in nomination first. Then McCarthy is said to have added that he could not have done it for Robert Kennedy but would do it for Ted. This gratuitous remark froze Smith—and probably right there killed any outside chance there might have been for an arrangement.

Steve phoned Teddy and passed on the information. Coolly, the thirty-six-year-old Senator appraised McCarthy's offer and the whole picture and, eventually, again said no. So soon after Bob's death, he didn't have the stomach for it as it stood. Looking at the matter practically, once McCarthy's name went before the convention, it would be hard for him to pull out and still deliver his troops. And Daley's interest and estimates of deliverable delegates always were suspicious; he had made it clear he thought the ticket needed Teddy in the second spot, and if Teddy became available for President and lost, it would be difficult then to say no to Vice President, which he flatly didn't want. Daley was reported later to have observed that Jack Kennedy could add, Bob Kennedy could add, but Ted Kennedy would have to learn. But it may have been that Ted could add all too well. The word was passed; it was all over. "If there is such a thing as a draft," Smith said later, "that was it. If Edward Kennedy wanted to lift a little finger he could have been the nominee. But he could foresee Nixon throwing his age and his family at him. And paramount was his determination that he didn't want to move on a wave of sympathy for his brother."

The rest was noisy, but predictable. The dovish Vietnam plank, supported by the old Robert Kennedy and McCarthy forces, lost by a 3-2 ratio and Humphrey was nominated on the first ballot. Ted Kennedy got 12¾ votes. Bill vanden Heuvel, in a last gesture of defiance, had threatened to put Lyndon Johnson's name in nomination, since it was his Vietnam policy and platform plank the nominee would have to defend. But he was talked out of it, he said, by White House emissaries. Of more interest now was the scene outside the Conrad Hilton downtown, where Mayor Daley's security forces broke and clubbed young demonstrators and newsmen, turning the convention itself into a battleground, with Daley as the central antagonist.

The Robert Kennedy presence was not yet over in Chicago, however. On the final night, a memorial film about him was shown in the convention hall, with a taped introduction by Ted Kennedy. "If my brother's life, and death, had one meaning above all others," he said, "it was this: that we should not hate but love one another, that our strength should not be used to create the conditions of oppression that lead to violence, but the conditions of justice that lead to peace." It was a message not necessarily intended for Chicago, but to many in the incendiary atmosphere that prevailed, it sounded apropos.

The thirty-two-minute film showed Robert Kennedy in every phase of his sixteen years of public service, and as it spanned over the years, a quiet enveloped the amphitheater. Suddenly there was the funeral of John Kennedy and the solitary days and weeks afterward, and then Robert's appearance before the previous Democratic national convention, four years earlier, to introduce a similar film about his brother and to take a party's tribute for that brother. Robert Kennedy's voice came on now in the Chicago amphitheater, as it had, softly, in Atlantic City, and the words took on new poignancy: "When he shall die,/Take him and cut him out in little stars,/And he will make the face of heaven so fine/That all the world will be in love with night/And pay no worship to the garish sun."

There were pictures of him climbing Mount Kennedy, campaigning for the Senate, and then for the Presidential nomination in Oregon and in California.

The voice came on again, this time with the vigor of the candidate. "This is a generous and compassionate country," Robert Kennedy was saying, in his urgent, pleading style. The convention hall broke into thunderous applause, and women began to dab their eyes, especially in the New York, California and Massachusetts delegations bunched together to the far left of the speaker's rostrum. "That's what I want this country to stand for," the familiar voice went on. "Not violence, not lawlessness, not disorder, but compassion, and love and peace. That's what this country should stand for, and that's what I intend to do if I am elected President." The applause built; even in death he got to the crowd more than many in life had been able to do through those bitter days in Chicago.

When it was over, the lights came up. For a moment it was quiet. Then the delegates rose and began to applaud—on and on and on, as if they were determined to match and exceed the tribute to Robert Kennedy and his brother at that previous Democratic convention. It went on, undirected and uncontrolled, for nearly five minutes.

Then, abruptly, what had been tribute became contest and tactic. House Majority Leader Carl Albert, the convention's permanent chairman, tried to gavel the convention to order, but his effort only spurred the supporters of Robert Kennedy to greater applause. Vanden Heuvel, tears streaming down his full face, and others began to sing "The Battle Hymn of the Republic," clapping in time, and soon the whole convention hall was filled with it, as Albert gaveled again and again for order. In the California delegation, a banner proclaiming BOBBY WE MISS YOU was unfurled, and in the last row of the gallery, some kids held another that said BOBBY, WE'LL SEEK YOUR NEWER WORLD. Although most delegations were in their seats now, the Kennedy loyalists in New York, California, Massachusetts, Wisconsin and a few others went on singing and applauding. The tears were gone and the mood was triumphant, defiant, Kenny O'Donnell, standing in the aisle, tapped his right foot in time, an impish smile on his Irish face. Rosey Grier, Rafer Johnson and actress Shirley MacLaine, in the California delegation, mouthed the words and kept clapping.

All the lost dreams, all the frustrations of the convention that might have nominated their champion, had been surfaced by the emotional film; the attempt to gavel away the tribute had been too much to bear, and now this one last fight was being waged—against the victorious Democratic establishment of Johnson, Humphrey and, yes, Daley, who had thrown in with them at the end. Finally, from the Illinois delegation, a signal was flashed to the galleries. Chicago ward-heelers and precinct captains, who had been given passes and orders to pack the hall, raised printed pro-Daley signs and began chanting "We love Daley!" The demonstration had been planned to counter abuse heaped on the mayor for the brutal conduct of his police, but now it was unleashed to put down the emotional outburst for Robert Kennedy.

For the next several minutes they went at it, further reducing the tribute to the slain Presidential candidate to a cheap shouting match. And finally, as the Daley chants continued, the Kennedyites abruptly stopped singing, turned away from the recent past, and began to chant the Camelot cry of the future: "We want Teddy! We want Teddy!" They were, without doubt, devoted to the memory of Robert Kennedy; but in the heat and the rancor of the moment, they resorted to the oversimplification, to the dynasty idea, and in so doing they demeaned the dead brother, and the surviving one.

Albert, determined to end the antics, recognized a Negro alderman from Chicago to deliver a short tribute to Martin Luther King and then, as the Kennedyites finally subsided to join a moment of silence for Dr. King, the convention chairman gaveled it away with an insensitivity that was incredible even for this most malodorous of political conventions, and went on with the business at hand.

Robert Kennedy's most fervent critics would have agreed that his memory deserved much better than all this from his friends as well as from his enemies. It was an unworthy way for a party to say good-bye to a man who—quite possibly—would have been its Presidential nominee, had he lived. Instead, his memorial was diminished into just another vehicle for the bitterness that infected his party in the convention hall, and that sent it divided and reeling into the general election of 1968.

The episode could be dismissed if it were not a commentary on the denigrating aspect of the Kennedy family legend, on the way it can rob the individual member of his unique achievements and identity. Robert Kennedy's singular promise as a man of reconciliation in a time of racial polarization, and as a charismatic personality whose passionate pursuit of this goal almost alone on the American political scene galvanized others, was blurred in the rush to get on with the restoration of Camelot, with the fulfillment of the legend. Even before the memory of Robert Kennedy had cooled, the faithful were turning to the last surviving brother as, in his own words, he picked up "a fallen standard" and began to carry it forward.

It will be unfortunate if for this reason the last campaign of Robert Kennedy is viewed simply as a kind of middle chapter in the great political family epic of our time. Though it may well have started as an elaborately rationalized attempt to restore John Kennedy's years, it became more than that. It brought forth a man of more obvious sensitivity to human suffering than John Kennedy was; it enabled that man to pull himself out of the shadow of his hero-brother and to imprint a unique and lasting identity of his own on millions of his fellow Americans; it was, above all, a personal tragedy not simply in the way Robert Kennedy died, but in the way he entered upon his final adventure and where it took him in his own heart and mind.

Like a hero in the Greek plays from which he was so fond of quoting to college audiences, he made one colossal error of judgment that set his fate, and his efforts to overcome that fateful mistake determined the nature and the course of his candidacy and his last campaign.

In the fall of 1967, before Senator McCarthy by default became the embodiment of the national discontent, Robert Kennedy was the almost classic example of the right man at the right time. He was ideally suited by temperament, compassion, emotional appeal, age, appearance and political skills to lead the effort of the nation's young activists to "turn the country around." At the same time, he was thoroughly rooted in the ways and the attitudes of the old political leaders and was both willing and able to deal with them on their terms. The charisma that enabled him to build his have-not constituency was not lost on the old politicians; had he added the young activists to the constituency by an early entry into an anti-Johnson, antiwar candidacy, he would have been exceedingly difficult to stop.

The Kennedy mystique, which without doubt was a major element in his charisma, worked not only on the blacks and the ethnics, but also on the old politicians. But the split in the antiwar ranks, resulting from the parallel Kennedy-McCarthy candidacies, prevented Kennedy from using the primaries to prove a public groundswell for him. Had he made McCarthy's candidacy unnecessary by himself challenging President Johnson in New Hampshire, the issue of "ruthlessness" that he insisted had kept him out likely would not have been raised so forcefully when he did elect to run. In the prevailing climate of deep public resentment toward Johnson, Kennedy's early entry might well have produced the effect that occurred in Kansas and California when he did become a candidate—an overwhelming outpouring of emotion toward the white knight come at last to slay the dragon. Had he been "ruthless" against the overbearing Johnson, he might have been acclaimed for it; instead he was "ruthless" against the quiet underdog, McCarthy, and he was castigated for it.

The problems of public men have a way of being reduced to shorthand. Thus it was for Robert Kennedy with that one word "ruthless." For all the humor about it in his last campaign, it was in a broad sense his undoing. Robert Kennedy was, in late 1967 and thereafter, a captive of his own past excesses, real or imagined, in much the same way Richard Nixon was in 1960. Then, Nixon—laboring under the shorthand label of "Tricky Dick"—campaigned inoffensively in an effort to counteract his own image of opportunism. In his first television debate, he sought to project an image of sweet reasonableness and was demolished by an aggressive and precise John Kennedy. In 1967, Robert Kennedy similarly held back from grasping a moment of huge political opportunity out of fear he would touch that most sensitive public nerve about himself. In so doing, he surrendered his undisputed leadership of the dissent.

The irony of it all was that, for all the talk of "ruthlessness," Robert Kennedy, like his late brother, was an extremely cautious politician. He liked to "walk around" problems, as one of his close aides once put it, and to exhaust his options before making a move—in much the way, in fact, that Lyndon Johnson did. This caution, coupled with Kennedy's misjudgment of the depth of the public sentiment against Johnson that he himself embodied, reined in Kennedy's own anti-Johnson impulses. It was only when his other emotions—his deep concern about the Vietnam War and about racial turmoil at home—pushed political judgment and caution aside that he acted. By then, circumstances were such that he only compounded the "ruthless" image and faced a political challenge, because of McCarthy's presence, far more difficult than it would have been had he run at the outset.

Even when Robert Kennedy did plunge into the race, his special label inevitably plagued him. Schooled in the cutthroat arena of politics, he practiced the art to the hilt, justified in his own mind by the rightness of his cause. As a man who thought in terms of moral purpose—he called war and bigotry "indecent" and meant it—he sought to strike a blow at the national conscience. In this effort his political schooling led him to use what weapons were necessary, such as the conservative rhetoric in rural Indiana. And in his passion, he sometimes failed to see that his determination and single-mindedness might be taken by others as "ruthlessness." For those closest to him, exposed daily to the quiet gentleness of the personal man, the label seemed incredibly inaccurate and unfair; but those who saw him only at a distance, who were his antagonists, could not be blamed for disagreeing. Robert Kennedy was obliged to become a public figure but he remained in most ways a private man; he gave his hands and his voice and his life to the public; his warmth, his soft, almost playful good humor he kept mostly for his family and those others closest to him. One of the most hopeful developments as he campaigned, however, was that the private man was blending into the public man; in time, had he been given the time, the bulk of the nation might have come to know him as his intimates did. His special constituency—the have-nots—seemed already to have accepted him thus.

It was, to the end, perplexing what exactly made him so wildly popular among Negroes and other minority groups. When he toured the riot area in Washington, D.C., in April, right after Dr. King was shot, a Negro woman came up, looked twice and said: "Is that you?" And when he nodded, she took his hand and said, "I knew you'd be the first to come here, darling." Others spoke of the plight of the Negro, and proposed remedies, but no other white politician had nearly the degree of acceptability, let alone adulation, that Kennedy enjoyed. "It was not so much what he said," Dick Goodwin suggested after Kennedy's death, "but they detected the same intensity in him a lot of them had." Edward Kennedy, reflecting on the campaign afterward, made the same point: "People who have to live so much by emotions, who depend on their feelings, can see sincerity in others. He felt deeply about the things he talked about and he showed it. They could tell he meant it." The emotional reaction of the crowds in turn had a profound impact on Robert Kennedy, his brother said. "The campaign personalized and intensified his concern. It happens in campaigns. I saw it happen before, with my other brother."

Yet this intensity that drove Robert Kennedy and, if Ted Kennedy and Goodwin were right, gave him identification with black Americans, also seemed to frighten people, especially those who disagreed with him. "One of his problems," Goodwin said later, "was that he was too well defined. John Kennedy in 1960 was vague, but Bobby was sharply etched. He polarized people right off." Frank Mankiewicz, writing in *Look* magazine, said, "He was an activist, and those who disliked him were not concerned that if he got power, he wouldn't do what he said; they feared that indeed he would."

Kennedy's intensity, his emotionalism, often was of such a degree that it swept aside both pragmatism and prudence. His entry into the Presidential race, after he had lost the clear leadership of the dissent and had to struggle with McCarthy for it, certainly was not at the time the act of a pragmatic politician. While it is true McCarthy's showing in New Hampshire underscored Johnson's vulnerability, it also complicated Kennedy's task immensely by splitting the anti-Johnson vote.

There were times when Kennedy's emotions drove him to actions that offered no political reward at all. His obsession with the plight of American Indians is the best example; the Indian vote was next to nothing, and there wasn't even much sympathy or awareness among the general electorate that there was an Indian problem. Yet Kennedy missed no opportunity to address himself to that concern, to the point where it became a laughing matter to nearly everyone in his entourage but him. "He became the champion of the poor, the blacks, the Indians." Mary McGrory wrote after his death. "Although possessed of advantages, he felt himself as one with the disadvantaged. He had been dispossessed of his treasure, his brother." It was that, and more. Like other rich men who have toiled in public service, he came to regard the rich as the enemy, and he despised the selfishness and lack of compassion he found among so many of them. He railed against the super-rich who paid little or no taxes. He offered a tax reform plan during the campaign to attack them. Inherent in his tirades against college kids who defended the student draft deferment in time of war was the knowledge that many of them came from well-off families that gave them everything but a sense of social responsibility. "He identified with people who hurt," Fred Dutton said after his death. "Maybe it was because he hurt."

In the years after John Kennedy's death, Robert Kennedy became the subject of unprecedented public adulation, heaped on him no doubt out of the depths of national sympathy and guilt that flowed from the first Kennedy assassination. In the eyes of most Americans, he entered the race for the Presidency in 1968 as, above all else, his brother's brother, whether he cared to admit it or not. But he left it as his own man to a degree that few appreciated. He never could be sure, even at the end, how much of the acclaim he received was directed toward his brother's memory and a voiceless yearning to restore the aura of the New Frontier, and how much toward Robert Kennedy. Nor could that acclaim ever really be sorted out. But as his campaign unfolded, and as he recognized that millions identified him as an instrument of racial reconciliation, he accepted the role, persuaded himself he might fill it and use it to mend the country's deep divisions. He always had the passion; now he had the purpose, and the conjunction made his practice of all the arts of politics more defensible and compelling to him. In the end, he was driven not by the desire to depose Lyndon Johnson, for that no longer was necessary, nor simply to put "the Kennedys" back in the White House. His compulsion, after all the days and weeks and months of wild adulation and the hands reaching out, was to somehow harness this outpouring of emotion toward him into a new sense of national purpose, looking inward to the needs of the have-nots in the affluent American society.

He was, in this particular time, the one man who could stir masses of people. When he struck at the consciences of people he triggered compassion in some, self-guilt and anger in others, but he moved them. Not only did he make students think twice about draft deferments; he made whites think twice about racial injustice and the affluent about the poor. When he failed to strike this note of concern, as in Oregon, he failed politically. But he was a candidate in a time when the country as a whole, unlike Oregon, felt it was indeed in trouble. There probably was never a more inept, ill-timed political slogan than Hubert Humphrey's "politics of happiness and joy," and Kennedy slashed it to pieces with his emotional chronicling of the nation's travail.

In a transitional period in American politics, when old coalitions were crumbling, Kennedy's knowledge of the country's mood, of the new political forces at work and of the old politics equipped him uniquely to practice transition politics. He was in harmony with the anti-establishment resentment toward bureaucratic, impersonal government; he had the appeal that could have stimulated massive grass-roots activity toward what the young activists liked to call participatory democracy. Yet he knew too that the old power brokers still held control, and he knew both the way to get to the back rooms and what to do when he got there.

It was this pragmatic ambivalence that so alienated the young activists, yet it was also the posture that made his chances of success so much greater than McCarthy's ever were, given the latter's disdainful attitude toward the old politics. The old pols didn't like Kennedy shaking the system, but they knew that he understood that system and they could survive, with adjustments, under a Robert Kennedy Presidency. McCarthy, on the other hand, was at war with them, and they didn't know what he would bring. The year 1968 marked a breakthrough for the new politics, but only that; the old politics hung on. Robert Kennedy might have achieved transition; in his absence there was, instead, chaos.

Had Robert Kennedy lived, the post-primary months preceding the convention would have marked an assault on the Democratic structure it had not seen in many years. Whereas McCarthy wasted these months in relative ineffectiveness, Kennedy already had a battle plan at the time of his death that would have allowed no letup from his exhaustive schedule to win the primaries. A strategy meeting was to be held the morning after the California primary, at which latest delegate counts were to be presented and key nonprimary states targeted for the onslaught. But first there was to be an all-out, two-week drive by Kennedy in his own faction-ridden New York State to nail down the bulk of its 190 delegate votes. As already noted, delegate-selecting primaries were to be held on June 18 in all forty-one Congressional districts in the state, with three delegates to be elected from each, and the rest at a state convention. In the wake of Kennedy's death, McCarthy made large inroads, but Steve Smith estimated later that Kennedy would have controlled all but 25 or 30 of the 190.

There was to be an intensive television effort blanketing the country, including two major TV documentaries on the primary campaign and short spots inserted repeatedly before the mass-audience shows, much as Richard Nixon did in the fall campaign. Kennedy was to barnstorm extensively in the populous East, in an attempt to spark the same excitement that occurred in all the primary states—all in the Midwest and Far West—in which he had campaigned. Even the possibility of a late primary challenge to Humphrey in some eastern state was being explored, to see whether there was any legal way to have a primary opened at that late date. In any event, there was to be hot pursuit of the Vice President by Kennedy wherever Humphrey went, in an effort to convert the preconvention period into a "national primary"—at least in the sense of discussion of the issues. There was to be saturation campaigning in all the major big-city ghettoes. In each, the plan was for Kennedy to go into the streets to demonstrate his vote appeal, and then to pay "courtesy calls" on the resident power brokers like Mayor Daley, to apply the heat. Daley's interest in Ted Kennedy in Chicago suggests that tactic may have paid off. Robert Kennedy is said to have observed once that "Dick Daley is the ball game," and if that was so, this approach was aimed directly at him. Finally, to cap all this, a trip abroad was planned for Kennedy, probably to Italy, Poland and West Germany, with a likely visit to West Berlin and John F. Kennedy Platz, to show what a Kennedy could do for sagging American prestige in Europe.

While Kennedy thus was occupied demonstrating his popularity at home and abroad, all the old hands would have been working the nonprimary states, pressing party leaders who were favorably inclined to take the leap, exhorting those who might be ready to go Humphrey's way to hold off. If Humphrey could be stopped on the first ballot, most in the Kennedy camp believed, the nomination would be theirs.

What was to be done about McCarthy was less clear, and even after all the weeks of dealing with his candidacy in the primaries, no headway had been made at arriving at an accommodation. Kennedy obviously had hoped that after his California victory McCarthyites would switch, and according to Allard Lowenstein later, that movement was possible, at least in New York. McCarthy himself said that there had been no talk between him and Kennedy of any accommodation, but that something might have been worked out at the convention.

Even if all the preconvention activity had gone well, Robert Kennedy would have faced long odds in the Chicago convention. But he might have pulled it off. In the utter madness of Chicago, the desperate search for an alternative to Humphrey certainly would have focused on Robert Kennedy. It would have been ironic if this young man, to many Democrats the symbol of party divisiveness, had been seized as the candidate of conciliation. Yet his demonstrated strength as a conciliator between whites and blacks, and his pursuits of a somewhat middle course on Vietnam between Humphrey and McCarthy, could have persuaded a convention reeling from political masochism to turn to him. Some of the Kennedy insiders had no doubts in their own minds what would have happened. "You have to presume," Steve Smith said later, "that Robert would have made it." And in a campaign against Richard Nixon, in which the winner would be chosen in the cities and the streets where his special constituency could be brought directly to bear, Kennedy might well have become what millions thought he someday inevitably would be—President of the United States. And in a Kennedy-Nixon campaign, with Kennedy as the antiwar candidate, he might at last have had back with him the issue-oriented young shakers whom he had lost in New Hampshire, and whose disenchantment with him had been one of his greatest sorrows.

All this is, of course, guesswork. Still, the prospect compounds the tragedy. The assassination of any Presidential candidate, particularly one so young, is tragic enough. Deepening it, in the case of Robert Kennedy, is the fact that he died with many of those who should have supported him most—the young activists—still turned against him. He might have won the nomination in 1968 and the Presidency; but in this time, with many of these young people whom he regarded to be the nation's hope of the future, he was the anti-hero, the spoiler whose personal ambition destroyed their chance, through McCarthy, to remake political America.

Perhaps later, as history gives its cold perspective to 1968 and to why Robert Kennedy followed the course he did, and as the activist young of today mature, they will come to see him as the transitional candidate that a time of political transition required. In what history says about Robert Kennedy, in fact, much depends on where those who rejected him are able to take the country in future years, and on how much credit is assigned to him for encouraging a national climate of dissent and of reconciliation in which they can bring change—and, as he hoped and believed, peace at home and abroad.

As Robert Kennedy found his own purpose and identity through the pursuit of his older brother's unfinished business, so in turn may Edward Kennedy. And it may be that in the future as in the past, Robert Kennedy's place in history, for all his individuality, will be determined by the fortunes and achievements of his brother. But for those to whom he was much more than his brother's brother, for those to whom he was the most electric political personality of his time and place, Robert Francis Kennedy will be remembered not merely as one of a family triumvirate, but as one man who in his own moment in history moved people, and was moved by them, to a condition of hope in a time of national disillusionment.

Epilogue

The Legacy

IN THE TWENTY years since the death of Robert Kennedy, the United States has seen a sea change in its politics. In June 1968, when Kennedy's own career was ended, conservatism was thought to be a rejected philosophy. It had, after all, been delivered a body blow by the electorate in the resounding defeat of Barry Goldwater just four years earlier. At the time, Ronald Reagan, in the second year of his governorship of California, was still regarded as a political aberration, a judgment seemingly confirmed by the failure of his halfhearted grasp for the Republican presidential nomination that was soon to go to Richard Nixon.

Nixon's narrow victory over Hubert Humphrey in November 1968 brought no wholesale lurch to the right. Nixon did attempt to undercut or kill certain liberal Democratic programs, most notably Lyndon Johnson's "war on poverty." But his election did not signal the kind of sharp and distinct break with the New Deal, and its successive mutations, that was to come more than a decade later under Reagan. The Republicans under Nixon mouthed all the customary clichés about the hated "welfare state," but they stopped far short of repealing it.

At the same time, the growing public unrest over the war in Vietnam, which had driven the candidacies of both Robert Kennedy and Eugene McCarthy and had persuaded Johnson against seeking reelection, pressured Nixon as well. After further American escalations of the war failed to achieve military success, Nixon was obliged in time to scale back. And in April 1975, with Gerald Ford now in the White House, the disaster that Kennedy and McCarthy had feared took place—America's first lost war, at a terrible cost in lives and American prestige.

On top of that foreign-policy catastrophe came the twin spectacle of a Republican vice-president and then a Republican president being driven from office in disgrace. This Republican embarrassment lulled the Democratic Party into a false sense of political security. In 1976 the election of Jimmy Carter, based heavily on the voters' hunger for a trustworthy leader after the nightmare of Watergate and Ford's subsequent pardon of Nixon, masked a growing middle-class unease in the country—and the emergence of the conservative movement from the ashes of the Goldwater defeat. Building upon this unease, and discontent with Carter, Reagan engineered his election in 1980.

Only now was the sharp change in America's political direction clearly perceived. Reagan in effect declared war on the basic social philosophy that had guided the country since the days of Franklin D. Roosevelt—enthusiastically pursued under most successive Democratic presidents, reluctantly under the two Republicans in the White House in that period of nearly half a century. With Republican control of the Senate during his first six years, Reagan was able to arrest or even roll back many of the social-welfare gains of that largely Democratic half century, until the voters again restored Democratic control of both houses of Congress in 1986.

The America of 1988, two decades after Robert Kennedy's death, probably bears little resemblance to the America that might have been had he lived and been elected president in November 1968. The narrowness of Humphrey's loss to Nixon, even after the divisive Democratic convention in Chicago, suggested that the prospect of a Kennedy victory over Nixon would have been good. And without Nixon, it is reasonable to conjecture that American involvement in Vietnam would have been terminated much sooner and in a much less ignominious fashion. Nor would there have been a Spiro Agnew vice-presidency and resignation, a Watergate, and the constitutional crisis that led to Nixon's resignation. "There wouldn't have been the enormous deterioration in public confidence in government," as Peter Edelman says.

Even without these national calamities, however, there is no firm assurance that Robert Kennedy, if elected, would have been more successful in ending the Vietnam War, in maintaining the social-welfare agenda with which his party was so long identified, and in rebutting the turn to the right that culminated in the Reagan Revolution. Nevertheless, there is some reason to consider that a Robert Kennedy presidency would have pursued an end to the war more urgently and would have forced this country to face its considerable domestic ills more directly than did the Nixon presidency.

Kennedy's close association with both the antiwar protest and civil-rights movements, the two paramount activist forces of his time, promised more realistic policies abroad and at home than those of Nixon, which permitted the war to drag on for nearly seven years after Kennedy's death and allowed racial disharmony to fester. A major question would have been whether he would have been able to diffuse the growing dissatisfactions of the middle class on which the conservative movement fed throughout the 1970s, culminating in Reagan's landslide election.

But such speculation is idle. Robert Kennedy did not live, was not elected president, did not end the Vietnam War, and did not detour the country's rightward shift that eventually crippled his party. Indeed, there is considerable doubt among even some of his most loyal associates of the 1968 campaign that he would have been the Democratic presidential nominee had he lived.

According to Richard Goodwin, who discussed his prospects with Kennedy in the privacy of his hotel bathroom moments before he went downstairs to make his speech proclaiming victory in the California primary, the candidate rated his nomination chances as no better than even. Kennedy and Goodwin agreed that too many delegates not subject to primary results were controlled by governors and other establishment political leaders beholden to Johnson and Humphrey. There was talk, Goodwin says now, but not much hope of trying to get McCarthy to withdraw "because Bobby didn't want to be chasing around New York State [site of the next primary] getting delegates against McCarthy, while Humphrey was consolidating his position. The news of the progress Humphrey was making [elsewhere] wasn't encouraging."

At the time of this discussion, the incomplete California results suggested that Kennedy had won the state so clearly that a discouraged or embarrassed McCarthy might indeed bow out. Instead, Kennedy's victory was narrow and McCarthy wasn't budging. And then there was Johnson to contend with. "I'm convinced," Goodwin says, "that Johnson would have done anything in his power, which would not have been inconsiderable, to prevent Kennedy from being nominated."

Others also express doubt that he would have won the nomination. Among them is George McGovern, who served as a stand-in behind whom the Kennedy delegates could rally at the convention. Kennedy's close call in California, and the establishment support for Humphrey, McGovern says today, probably would have been too much for him to overcome. However, he adds, had Kennedy survived he believes he would eventually have been elected president, possibly in 1976 when Republican misfortunes opened the way for a Democrat, and the party was looking for strong leadership. And as president, McGovern says, Kennedy might well have spared the country the Reagan Revolution.

Most others I have talked to recently, however, believe Robert Kennedy would have been nominated and elected in 1968. One of them is Tom Hayden, the onetime student radical, neighborhood organizer, and now California state assemblyman. He says that whether myth or probability, "If you believe that [Kennedy would have gone to the White House], his death becomes one of the central events of your life. That single event was the death of hope for peace in Vietnam for five, six or seven years, and the death of political hope for many people, myself included. It confirmed a paranoid view of American society for a lot of people. In a lot of places I've met burned-out Robert Kennedy workers. It took them maybe ten years or longer to reinvolve themselves. We worked for the integration of the 1960s generation into the process, and instead we got the opposite, and many of them are still looking."

Hayden says Robert Kennedy left behind "a different kind of Camelot [from John Kennedy's]—a Camelot of conscience. The first Camelot was more an image of bright and beautiful and positive people. The second, because of all that happened between 1963 and 1968, left an image of utopian struggle against overwhelming odds that failed and left a sense of unfulfilled hope . . .

"Bobby represented a commitment to the poor and a toughness, and what he was putting together was not a traditional Democratic coalition. Humphrey had the AFL; Kennedy had Cesar Chavez fasting in Delano, Indians on the reservations, street people in Bedford-Stuyvesant—the unorganized poor. The poor were an existential focus for him. Youth fascinated him, and they were also outside the old Democratic coalition. He was developing his own coalition, using youth as the shock troops. He had a unique opportunity to keep the traditional Democrats in the fold and convert the students. When he died it was too late to work the magic—and it was magic, that force of personality, because the two [factions] didn't have that much in common. One wing of the Kennedy coalition became absentee and another went to Nixon."

The notion that Robert Kennedy as president might have detoured the conservative movement that flowered under Ronald Reagan is a questionable one at best. Arthur M. Schlesinger, Jr.'s basic premise in *Cycles of American History* holds that the political pendulum has swung from left to right and back again in roughly chartable cycles throughout the life of the Republic, with events and personalities shortening or stretching out one cycle or another at times but not altering the essential pattern. Ted Sorensen agrees with Schlesinger that the pendulum "was bound to swing" to conservatism eventually whether Kennedy had been elected president or not. Goodwin argues, however, that the "threads of liberal legislation" advanced by John Kennedy and Lyndon Johnson, coping with urban unrest, poverty, and civil rights, were dropped in 1965 because of the demands of the Vietnam War. "Bobby would have picked up those threads and gone on," Goodwin suggests. "He might well have been able to sustain that liberal impulse for quite a while."

To contemplate what might have been, however, is an empty and frustrating exercise. It is more constructive, and more satisfying, to reflect on the legacy Robert Kennedy did leave, not only his tangible achievements but also the ambition and spirit he demonstrated and inspired in others. For many of the individuals who shared his last great political adventure, there has been a feeling of opportunity lost as they have gone on to other pursuits. They have remained guided by memories of his rare combination of compassion and toughness, his "passion for public service," in McGovern's words. But they have come away with a heavy sense that time and circumstance have robbed them and the country of a rare leadership it is not likely to discover again for a long time.

Some, like John Seigenthaler, lament the penchant of today's politicians to chart their directions by reading the public-opinion polls and shaping their policies and rhetoric to appeal to the majority. Kennedy, Seigenthaler says, read polls as a challenge to redirect public opinion, not coddle it. "He had a moral compass," he says, "and he sought to direct debate and dialogue in a direction that raised the consciousness, and the conscience, of the electorate. That's what I miss."

That compulsion to take on the comfortable and the complacent, often in their own dens, is what many of Robert Kennedy's political and personal intimates best remember about him. His eldest children, Kathleen Kennedy Townsend and Representative Joseph P. Kennedy III, both emphasize this trait in reminiscing about their father. "When he spoke he wasn't willing to gloss over differences," Kathleen says. "He was willing to take on the students who were all against [the war in] Vietnam, but were taking student deferments." And from his congressman son: "There was a certain spontaneity that would allow him to speak and act from his gut, that created a certain honesty. And whether that honesty drew him into fights that would make people angry, or whether that honesty drew him into fights that ended up making people admire him, it didn't make any difference. Because his pursuit was kind of an honest assessment of what was going on, mixed with a sense of trying to get the job done, rather than just appealing to what was the most liberal position or conservative position."

Jeff Greenfield puts it more directly: "The bullshit quotient by Bobby was extraordinarily low. He almost delighted in going before an audience and telling them where he disagreed with them—students on Vietnam, liberals on crime." Goodwin says: "He made people feel guilty for not doing." Frank Mankiewicz notes: "People knew he was a tough cop. He was a fighter who took advantage of his special place to get things done." And Ed Guthman: "He had the quality of seeing something and having to do something about it, whether it was getting a bird off a ledge, dealing with the Washington public schools, Indians on the reservations, Bedford-Stuyvesant. And he got people of talent and imagination—people like Eli Jacobs, a Republican, not only to raise a hell of a lot of money for Bedford-Stuyvesant but to give up a year to it."

This compulsion to grapple directly and personally with problems, Joe Kennedy says, was one thing that set his father apart from John Kennedy. "President Kennedy, I have a hard time relating to his tremendous breadth of capabilities, which seemed to come, at least when you look at it now, so naturally; to get along with everybody and always be kind of the fellow who was able to rise above the fray and have everybody work toward [his solution]. Whereas I think Daddy was the type of person who jumped into the middle of the fray and fought, and made enemies, but made friends. If you do that for a while, people think very harshly of you. If you do that for a very long while, you begin to earn respect. Here was a guy who basically came from an Irish Democratic family, who took on the labor unions yet was fighting for the right people. . . . You can make all the good speeches, you can sound all the right notes, but [it doesn't matter] unless you let your gut take you, which is what I think Daddy did. And he hit it so hard, on everything."

And then, finally, there was the special connection Robert Kennedy was able to make with people who hurt. "The thing that occurs to anybody who goes through a tremendous loss when a death occurs," Joe Kennedy says, "is there's a tremendous sadness. And out of that sadness comes a certain kind of compassion for people who are going through a lot of hurt and pain that enabled him to hook up a little bit better with people who were hurt by the policies that were taking place at that time, and to speak more freely from his gut, because the sense of hurt was so much more dramatic." But the notion that there was some huge metamorphosis in the man, he says, "was something identified by people who got to know him when he became more public himself. But the fact is he always had an ability to be very tough but also to be very gentle in almost the same moment."

And Kathleen Townsend: "He understood suffering, and a lot of people were suffering. In the way John Kennedy in running for president tried to raise the level of people's ideas, my father talked about suffering in a way that resonated deeply in people's hearts, until they felt something larger than themselves. And it was not just talking about getting things done. Like Bedford-Stuyvesant, he got them done. And he didn't think words like liberal and conservative were very appropriate. Maybe that's why liberals were always very uncomfortable with my father."

Melody Miller, the receptionist in Robert Kennedy's Senate office who after his death went to work for Ted Kennedy and eventually became his press secretary, recalls a television cameraman in the 1968 campaign who said: "I like this guy. He's got a perpetual sense of outrage against social injustice." She recalls too how Robert Kennedy would return to his Senate office after a visit among poor blacks in the South, go to his desk, and personally start soliciting friends to send food and clothing. "He introduced legislation too," she says, "but he was frustrated with the slow legislative process."

Fred Dutton, the Washington lawyer who was Robert Kennedy's most trusted adviser on the campaign trail, says he is amazed at the durability of Kennedy's appeal, considering he never won his party's nomination, let alone was elected president. Along with the idealists "who say that if Bob Kennedy had been elected president the world would be a Utopia today," Dutton says, are those who remember him more as just "a gutsy guy, a hard-nosed guy," whose appeal was in his striving. "It was summed up in Bobby's line, 'We can do better.' "

So what, all these years later, has been the real legacy of Robert Kennedy? There is a strange ambivalence in the answers. On the one hand, those who were closest to him in his political life talk about him as a man of action even more than a man of words. Yet they find themselves dwelling on the message the country lost in his premature death, much more than on the tangible achievements wrought by his actions. That, perhaps, is in the very nature of an early death that robs promise.

David Broder suggests that Kennedy "has been a very important symbol in the battle for social justice, civil rights, and all of the other issues with which he was identified. His legacy is much more as a symbol for causes than a direct political impact on the country today." Kennedy, Broder says, touched "something in the spirit of the times, but the sixties were different. Maybe others can re-create the rhetoric, but they can't re-create the atmosphere that existed then." And, he observes, not an overwhelming number of people who were part of the Robert Kennedy presidential campaign of 1968 remain active in the front lines of American politics today.

Many of the prominent participants in that campaign have gone either into the practice of law with political offshoots, such as Peter Edelman, Ted Sorensen, Fred Dutton, Joe Dolan, Adam Walinsky, or into some aspect of journalism—such as Ed Guthman, Jeff Greenfield, Frank Mankiewicz, Pierre Salinger, John Seigenthaler.

There are some, however, who have stayed on the front lines, or have come back after a time. One such is Brock Adams, a U.S. Attorney in Seattle in Robert Kennedy's Justice Department, one of the few congressmen to endorse his presidential candidacy in 1968 and campaign for him, and now Washington State's junior senator. Adams was perhaps the prime long-shot Democratic winner in the 1986 elections that regained Senate control for the party. Kennedy's determination helped inspire his own decision to run, he says, and to persevere when his campaign was struggling, without money, and dismissed by others as hopeless. "Once I made the decision," he says, like Robert Kennedy, "I never looked back."

There are a great many other individuals, many in the generation that was just becoming politically aware at the time of the 1968 campaign, who identify Robert Kennedy as the prime motivator in their decision to involve themselves intensely in politics, and in keeping them at it regardless of the setbacks and disappointments delivered upon them by the onrushing Reagan Revolution.

One is civil-rights activist John Lewis, now a Georgia congressman. "I probably wouldn't be where I am if it weren't for Robert Kennedy," he says. "More than any other politician of my time, along with his brother, he inspired a whole breed of young people to get into politics. I saw in his campaign a sense of daring, a sense of courage that said if you really believe in something, you've got to fight for it. He had the ability to go straight to the heart of an issue, like civil rights or Vietnam. And his tie with blacks, there was nothing false about it. Black voters who with others would be unresponsive sensed something about Bobby Kennedy that was different from any other politician. He was able to communicate not only with his lips but with his heart and guts. Those of us who came through the civil-rights movement dealt with him on a different level. We knew him as a fighter, standing up for the little man in the South and in South Africa. Today in black churches and homes all over the South and the country, they have a frame with the pictures of Martin Luther King, John Kennedy, and Robert Kennedy."

Others whose political awareness was awakened by Robert Kennedy have become important figures in contemporary presidential campaigns. One is Paul Tully, who dropped out of Yale for two years and was steered into Kennedy's Senate campaign of 1964 by Allard Lowenstein, then active in civil-rights and antiwar protests on campuses, including Yale.

"At first, while Kennedy had a sense of energy and action, publicly he was not a good candidate," Tully says. "He had a lot of views, strongly held, but they were sketchy. But he seemed to be learning. He was the only candidate I worked for up to that time who you watched grow daily."

After Robert Kennedy was in the Senate, Tully went back to Yale and after graduation was drawn into politics again in the fall of 1967, when Lowenstein and Curtis Gans were shopping around for a candidate to challenge Lyndon Johnson for the Democratic presidential nomination. When Kennedy declined to run and Eugene McCarthy became the "Dump Johnson" candidate, Tully signed on, working in Connecticut and New Hampshire; but he, among other student activists, became disenchanted with McCarthy's focus.

"McCarthy was not about winning, he was about a statement, almost an aesthetic statement," Tully says. "He drew a group of terrifically qualified people around him, motivated toward urban, ethnic problems, and the state of the dispossessed. But McCarthy was more concerned with foreign policy and what was wrong with the [political] process. Kennedy conveyed the idea that you were playing for real stakes, that you were playing for the outcome, not just unseating Johnson." (McCarthy, asked for his view of Robert Kennedy's legacy, referred not to his presidential campaign, which intruded on McCarthy's own effort, but to the Justice Department under Kennedy, which had pursued some important cases, he said.)

Increasingly put off by McCarthy's approach in 1968, Tully and others switched to Robert Kennedy soon after he entered the race. "Kennedy was the first politician who conveyed to a lot of us that what you did possibly would help change the real world," Tully recalls now. "You didn't have to just sit there. Whether it was racism or the war or the rest of President Kennedy's agenda, he didn't leave the sense of being held back by dominant, corrupt institutions. You faced it; you fixed it. He was tough and he acted."

After Kennedy's death, Tully devoted himself to either organizing at the grass-roots level for progressive organizations or helping run major political campaigns. He headed Ted Kennedy's presidential effort in Iowa in 1980, was political director for Walter Mondale in 1984 and for Gary Hart until the middle of Hart's 1988 campaign, and then for Michael Dukakis until Tully became a casualty of the videotape-leaking incident that helped bring down the candidacy of Senator Joseph Biden in September 1987. Tully notes a host of community leaders has sprung up in cities across the country, fighting for the disadvantaged and powerless. "A whole generation came to learn that just because it wasn't easy you still do it," he says.

An even younger member of the later generation, similarly motivated by the Kennedy example, is John Emerson, a chief political lieutenant to Hart in 1984 and 1987 and, since then, active in rallying progressive Democrats in his home state of California. Emerson was in the ninth grade when Robert Kennedy was killed, and he insists his generation has been energized more by Kennedy than even Kennedy's own was.

"So many of the people who worked with him are still almost shell-shocked, so they haven't gotten back in," Emerson observes. "They were shattered, and for them no other human being can measure up to him. His real legacy is with people of my generation who became attuned to injustice in our society through Bobby Kennedy, his candidacy and the way he expressed things. In terms of what one man can do, Bobby Kennedy is much more real to people of my generation than John Kennedy ever was. We were too young. Whereas our elders were disillusioned [by his death], we weren't close enough to be shattered; instead, we were motivated to see what could be done. As a result of Bobby Kennedy, I absolutely refuse to accept that any political problem is not solvable. He would march into the most difficult situation and sit down with people and tell them there wasn't a problem that can't be solved with the right energy and by facing your anger and frustration. That made a lot of sense to our generation."

Emerson calls Kennedy's contribution "an attitudinal legacy—a commitment to eradicate injustice and a sense of confidence that even if we suffer major setbacks along the way, anything can be accomplished if we have the energy, commitment, and courage behind it." This attitudinal legacy, he says, has kept him going even after the disappointment of the Hart campaign. "I'm more determined than ever, because we're not going to be beaten down or let a situation that happened stop a greater cause."

Still another who personifies the Robert Kennedy legacy is Rick Allen, John Emerson's law-school roommate, a thirty-three-year-old Los Angeles businessman who was fourteen years old and handing out campaign leaflets for Robert

Kennedy in New York at the time of his death. Allen recalls lying on a small bed in his mother's bedroom, while she and his sister slept, listening to the California primary returns on a radio under his pillow when the news of the shooting came. "I was supposed to be asleep," he recalls now, "but I knew we had to do well in California to do well in New York. When I heard the news, I didn't want to wake up my mother and sister, so I just lay there and cried to myself."

Robert Kennedy, he says, was the political guidepost "for those of us too young to be affected by his brother's death. I remembered it as a national tragedy, but I remember Robert Kennedy's death as a personal tragedy." Allen talks of Kennedy's "muscular liberalism," which "demanded excellence but was compassionate toward those who couldn't keep up the pace. He stood for the idea that energy and talent can be applied to ultimately make things better. People today don't think leadership is possible, that anybody has the ability to change the world. I'm thankful to Robert Kennedy for not thinking that way. He said that individuals make a difference, and because violence in institutions is so destructive, that's why individuals have to become involved." Allen also worked in the Gary Hart campaign and returned to California, where he has been involved in plans for endowment of the Robert F. Kennedy Memorial Foundation.

Along with the conviction that Robert Kennedy left a legacy in the message that individuals can make a substantial difference, there is, however, a widespread disillusionment about the quality of leadership in his absence. "What he stood for," Ed Guthman says, "is still a powerful force in the country, but it just can't get its head up." Reflecting on the public mood in mid-1987, George McGovern said: "There's a hunger out there in the country for a strong, clear, forceful leader. I sensed that even in 1984. There's a vague uneasiness. If there was a commanding figure out there like Robert Kennedy now, I think he'd be on his way to the nomination."

Tom Hayden agrees. "In a certain sense, the country's always been ready for somebody like this," he says. "The country does go through cycles, but if Bobby Kennedy believed anything, he believed that you've got to create your own cycles, and go against the tide. That's what he did. He said it was time to shift the country, not wait."

Jeff Greenfield calls it a great irony that "twenty years after Robert Kennedy's death, the Democratic Party is coming around to the notion that big government and more programs is not the answer." The melding of opportunity and obligation as Kennedy preached, he says, "was much less likely to alienate the taxpaying middle class." He suggests that had Kennedy been able to accommodate traditional Democratic liberalism to this melding, the party's commitment to change "could have been modified before it became ossified."

Peter Edelman puts the same point another way. Kennedy, he says, "was the first person who had any power who began to try to grope toward a post–New Deal liberal position—a thoughtful, independent way different kinds of arrangements and levels of government could be used to solve problems. If he had lived, even if he had not been nominated and hadn't become old hat by now, or if there was a younger version of him, he would be a very relevant person right now." Edelman suggests that the country, having gone through a conservative cycle that has perhaps been cut short by events, "is groping our way back," as shown most recently by the rhetoric of several 1988 Democratic presidential candidates in support of activist government.

Another irony is that the Democratic Party does have a leader of demonstrable passion, compassion, and oratorical skill who does clearly carry on the legacy of Robert Kennedy. That is his brother Ted, who perhaps more than any other figure in public life has absorbed and sought to fulfill the mission of his brother in spirit, word, and action. Although Ted Kennedy once failed in a bid for the Democratic presidential nomination, and on five other occasions has declined to run, in his performance as a senator and party leader he has committed himself to issues and projects that are in the high-risk tradition of Robert Kennedy. The circumstances of his personal life, the tragedy on Chappaquiddick foremost, have kept him from regaining Camelot for the Kennedy legend. But he has been diligently at work, nevertheless, seeking the policy goals that motivated both of his brothers, but Robert particularly.

Those who remember Ted Kennedy when he first came upon the national political stage as a senator will recall that he was generally perceived to be much more like John than Robert. He was more outgoing personally, yet more cautious in his politics, as seen in his original skepticism about Robert challenging Lyndon Johnson. But since Robert's death. Ted too has challenged an incumbent Democratic president whose commitment he has questioned. And he has become a workhorse on the issues of greatest concern to the second brother, from Vietnam War protests to racial inequality at home and abroad.

Camelot, however, has not been retaken by Ted Kennedy, in the sense of restoring the family to national power, a development that might have reduced Robert to an interim player in a grand drama of dynasty resurrection. Accordingly, the speculation at the close of this book written twenty years ago that "it may be that in the future as in the past, Robert Kennedy's place in history, for all his individuality, will be determined by the fortunes and achievements of his brother," has not been borne out.

As noted, a whole generation of young Americans has grown into maturity with Robert Kennedy, not John, as its inspiration for political action and commitment. In the end, this man who during his brief life was often criticized and disliked for a coldness and, yes, ruthlessness in the pursuit of his brother's and his own political objectives, is remembered in death for the passion of his beliefs and the compassion of his actions, in his own right. Much of the bitterness toward him among the young who supported Gene McCarthy has faded as they have come to view him in a more historical framework, and have come to recognize in him an embodiment of their own better instincts.

It is interesting to note that the Robert F. Kennedy Journalism Awards, created by reporters who covered his last campaign and now administered by the Robert F. Kennedy Memorial Foundation, have become the second most-sought-after prizes in American journalism, behind only the Pulitzer Prizes. The journalism awards go for newspaper, magazine, radio, and television reports and commentaries on the plight of the disadvantaged, and in the twenty years since Robert Kennedy's death the entries have shed a revealing light on all manner of societal ills and victims. Also much honored are the RFK Human Rights Awards, made for the fourth year in 1987. Recipients have included political dissidents in South Africa, Poland, and South Korea. In 1986, Ethel and Ted Kennedy journeyed to Poland to present them.

The question of what the country might be like today had Robert Kennedy lived is one that has frustrated many who came to political awareness as a result of his brief, emotional dash for the presidency in 1968. As noted, some dropped out of politics in disillusionment after his death. Others, in Edward Kennedy's words, picked up the fallen standard and became foot soldiers, lieutenants, and even generals in the later presidential campaigns of George McGovern, Morris Udall, Ted Kennedy himself, Walter Mondale, Gary Hart, Michael Dukakis, Paul Simon, Jesse Jackson, and other liberal Democrats. Together they have become a cadre of highly motivated political troops, toiling in the spirit of Robert Kennedy.

The enthusiasm of these loyalists has not, however, been matched in the public at large. American presidential elections have been marked increasingly by voter apathy, in both campaign participation and at the ballot box. Ronald Reagan has demonstrated in the Republican Party the importance of a leadership that can arouse a sort of self-centered patriotism. But the Democratic Party since Robert Kennedy's death has been devoid of a politically credible voice that speaks with an urgency capable of moving people, black and white, selflessly as he did.

In some ways, the challenge is more difficult today than it was for Robert Kennedy twenty years ago, when the Vietnam War and racial unrest at home provided ready fuel for an emotional approach to the electorate. The complacency that has captured the country since then, encouraged by an administration that disparages the poor and the disadvantaged, may be too much for even the voice of another Robert Kennedy to overcome.

On this point, listen to a member of the post-Kennedy generation with particular insight—Joe Kennedy: "You go through the decade of the seventies; first you get the Watergate, then you get the mixed-up Ford presidency, then you get tremendous disappointment in the Carter presidency, then you've got Mr. Feelgood. We've got the feelgood politics of the eighties, and what we've seen as a result of that is that we now have the greatest disparity between the well-to-do and the poor. You see the shrinking of the middle class in this country, and yet the hurt and the pain is hidden through the mechanism of running up a trillion and a half dollars' worth of debt.

"So nobody's out there feeling the pain we all should be feeling if we don't want to pass on this debt to our children. You can stand up and make that speech today, but nobody feels the pain. . . . If you had to go in and sequester two hundred billion dollars a year and get this budget in balance today, the pain that would be felt out there I think would open up the gap for the kind of politics my father was talking about to spring forth again. Right now it just doesn't work because people aren't feeling the hurt. . . . Until these cuts really start to hit home, I don't think it's really going to hit people. . . . The poor blacks know about it, the Indians know about it, the Eskimos know about it, but the trouble is they're all isolated groups now."

In the same vein Joe Kennedy observes that the sight of the homeless on so many city streets today makes people "embarrassed that in America we have this ugly advertisement of what's really happening. So they want to clean that up. But there isn't any real burning desire to get rid of homelessness. . . . As far as getting at the gut level, root causes of this problem, nobody's speaking about it."

Clark Clifford, friend and counselor to Democratic presidents for half a century, suggests however that the Reagan tenure in an ironic way enhances the Robert Kennedy legacy. "Ronald Reagan has been almost the direct antithesis to Robert Kennedy in his approach to what Franklin D. Roosevelt called the ill-housed, ill-fed, and ill-clothed," Clifford says. "Bobby Kennedy had a real and genuine feel for these people and has had a continuing impact on attitudes toward them. His link to them has kept that concern alive during the Reagan administration. Unquestionably, the Reagan experience has made Bobby Kennedy's legacy shine more brightly through the gloom and darkness of this period, because the contrast is so dramatic that those who believed in his philosophy and values have been confirmed in that belief."

Arthur Schlesinger wrote in his Foreword to *Robert Kennedy and His Times* that although his "message of commitment to the desolate and disinherited is rarely sounded [anymore], the time will surely come again when the richest nation will overcome its indifference to the degradation of its citizens. When that time returns, Robert Kennedy's journey will, I believe, be seen to have exemplary value."

The fact that much of the country seems to have lost its capacity for outrage over the widespread neglect of the downtrodden, as well as over the abuses of power in high places, cannot be attributed to the absence of a single political leader, no matter how eloquent he may have been. Others, most recently and prominently Jesse Jackson, have sought to arouse such outrage with some success, but not of the breadth that Robert Kennedy achieved in reaching across racial and economic lines before and during his 1968 presidential campaign. The continuing need for such a voice in the America of the late 1980s measures the cost of his untimely death.

Acknowledgments

In the annals of Presidential campaigns, there may never have been another like the last campaign of Robert Kennedy. Begun late and ended prematurely, it moved across the landscape and the emotions of America at a feverish pace. A candidate and a professional team accustomed to practicing the art of winning elections in all its sophistications were thrust by events into one improvisation after another as the spring of discontent sped by. In six primary elections within four weeks, the candidate struggled to make up for lost time and lost opportunity; in the process, he sought to strike a responsive chord in enough voters and political leaders to shake the foundations of the Democratic Party and capture the Presidential nomination in a year when the incumbent had been expected to seek a second term.

When that feverish effort ended in Los Angeles, the odds still seemed great that the candidate would not succeed; as subsequent events unfolded, however, few were quite so certain as they had been. It is probably fair to say that before Robert Kennedy sought the Presidency, most Americans assumed he would occupy the White House someday. This book attempts to chronicle why he elected to hasten what appeared to so many to be the inevitable and how he attempted to implement that decision in a time of great political upheaval.

In the short time since Robert Kennedy's death it has been possible to provide this account only because many associates of Robert Kennedy and many who reported his activities have generously contributed their recollections. Among those in the first category are Senator Edward M. Kennedy, William Barry, John Douglas, Richard Drayne, Frederick G. Dutton, Peter Edelman, Richard N. Goodwin, Allard K. Lowenstein, Frank Mankiewicz, Bill D. Moyers, John Nolan, Angela Novello, Lawrence F. O'Brien, Kenneth O'Donnell, Pierre Salinger, Stephen E. Smith, Theodore C. Sorensen, Richard Tuck and William vanden Heuvel. Among those in the second group are Bruce Biossat, especially, and Walter Cronkite, Richard Dougherty, James Doyle, Rowland Evans, Richard Harwood, Paul Hope, Anthony Howard, Haynes Johnson, Joseph Kraft, Roger Mudd, Warren Rogers, Richard Stewart and Sander Vanocur.

To all these, I express my gratitude for their help in this effort to portray the struggle and the spirit that made the last campaign of Robert Kennedy a memorable experience for all who witnessed it. My thanks also to Arthur C. Fields, my editor at Putnam, for his constructive counsel and to my wife, Marian, for long, painstaking hours of clerical and editorial assistance and advice.

Jules Witcover
Washington, D.C.
October, 1968

Index